D0805522

SPRINGER-VERLAG

D-6900 Heidelberg 1
P. O. Box 105280
Telephone (06221) 487·1
Telex 04-61723

D-1000 Berlin 33
Heidelberger Platz 3
Telephone (030) 822001
Telex 01-83319

SPRINGER-VERLAG
NEW YORK INC.

175, Fifth Avenue
New York, N.Y. 10010
Telephone (212) 477-8200

STRUCTURE AND BONDING

Volume 38

With 31 Figures and 9 Tables

Springer-Verlag
Berlin Heidelberg New York 1979

G. A. Somorjai and M. A. Van Hove
Department of Chemistry, University of California,
Materials and Molecular Research Division,
Lawrence Berkeley Laboratory, Berkeley, CA 94720, USA

ISBN 3-540-09582-9 Springer-Verlag Berlin Heidelberg New York
ISBN 0-387-09582-9 Springer-Verlag New York Heidelberg Berlin

Library of Congress Catalog Card Number 67-11280

Typesetting: H. Charlesworth & Co Ltd, Huddersfield
Printing and bookbinding: Brühlsche Universitätsdruckerei, Gießen 2152/3140-543210

Adsorbed Monolayers on Solid Surfaces

Gabor A. Somorjai and Michael A.Van Hove

Contents

I Introduction

An atom or molecule that approaches the surface of a solid always experiences a net attractive potential[1]. As a result there is a finite probability that it is trapped on the surface and the phenomenon that we call adsorption occurs. Under the usual environmental conditions (about one atmosphere and 300 K and in the presence of oxygen, nitrogen, water vapor and assorted hydrocarbons) all solid surfaces are covered with a monolayer of adsorbate and the build-up of multiple adsorbate layers is often detectable. The constant presence of the adsorbate layer influences all the chemical, mechanical and electronic surface properties. Adhesion, lubrication, the onset of chemical corrosion or photoconductivity are just a few of the many macroscopic surface processes that are controlled by the various properties of a monolayer of adsorbates.

In this paper we shall review the various experimental parameters that can be used to characterize the adsorbate layer in the sub-monolayer to few-monolayers range. Then we shall discuss the principles of ordering of the adsorbate layer, since one of the most exciting observations of low energy electron diffraction studies is the predominance of ordering within these layers. We shall list the ordered absorbate layer structures and shall summarize what can be learned about the nature of their bonding from the available structural data. This will be done separately for the many (~1000) surfaces whose two-dimensional unit cells are known in terms of shape, size and orientation, and for the fewer (~100) surfaces for which additionally the contents of the unit cell are known (adsorption site, bond lengths, etc.). Many types of adsorption will be covered, including atomic and molecular adsorption, co-adsorption, metallic adsorbates, non-metallic adsorbates and organic adsorbates.

Reference

1. See, for example, Bardeen, J.: Phys. Rev. *58*, 727 (1940); Lennard-Jones, J. E.: Trans. Faraday Soc. *28*, 28 (1932)

II Principles of Monolayer Adsorption

Consider a uniform surface with a number n_0 of equivalent adsorption sites. The ratio of the number of adsorbed atoms or molecules, n, and n_0 is defined as the coverage, $\theta = n/n_0$. The coverage in the monolayer is usually less than or equal to unity for a uniform surface. For a heterogeneous surface that exhibits multiple binding sites, i.e., more than one site per substrate unit cell, small adsorbate atoms may build up coverages somewhat greater than unity. We shall, however, ignore this possibility for the present.

When adsorption occurs on the clean surface, heat is liberated during the formation of the surface bond. The heat of adsorption, ΔH_{ads}, associated with the layer of adsorbates reveals the strength of interaction between atoms and molecules in the monolayer and the surface on which they are adsorbed. These two macroscopic, experimentally measurable parameters, θ and ΔH_{ads}, usually well characterize the adsorbed monolayer and the form of their interdependence often reveals the nature of bonding in the adsorbed layer.

Atoms or molecules may impinge on a surface from the gas phase where they establish a surface concentration $[n_A]_s$ [molecules/cm²]. Let us assume that only one type of species of concentration $[n_A]_g$ [molecules/cm³] exists in the gas phase so that the adsorption process can be written as

$$A(\text{gas}) \underset{k'}{\overset{k}{\rightleftarrows}} A(\text{surface})$$

and the net rate of adsorption may be expressed as

$$F\left[\frac{\text{molecules}}{\text{cm}^2\text{s}}\right] = k[n_A]_g - k'[n_A]_s \tag{1}$$

where k and k′ are the rate constants for adsorption and desorption, respectively. Starting with a nearly clean surface, far from equilibrium, the rate of desorption may be taken as zero and Eq. (1) can be simplified to

$$F\left[\frac{\text{molecules}}{\text{cm}^2\text{s}}\right] = k[n_A]_g \tag{2}$$

where

$$k = \alpha\left(\frac{RT}{2\pi M_A}\right)^{1/2} \quad [\text{cm/s}],$$

α is the adsorption coefficient, M_A is the molecular weight of the impinging molecules, R is the gas constant and T is the temperature. The surface concentration $[n_A]_s$ [molecules/cm^2] under these conditions is the product of the incident flux, F, and the surface residence time τ [s]:

$$[n_A]_s = F\tau. \tag{3}$$

If the incident molecules stay on the surface long enough to achieve thermal equilibrium with the surface atoms, τ has a form of $\tau = \tau_0 \, e^{\Delta H_{ads}/RT}$ where τ_0 is related to the average vibrational frequency associated with the immobile adsorbate. The value of τ_0 may be markedly different if the adsorbate possesses one or two translational degrees of freedom along the surface.

The heat of adsorption that is defined as the binding energy of the adsorbed species, is always positive. Clearly, the larger is ΔH_{ads} and the lower is the temperature, T, the longer is the residence time. For a given incident flux, larger ΔH_{ads} and lower temperatures yield higher coverages. Substituting the vapor density by the pressure using the ideal gas law $[n_A]_g = N_A P/RT$ (where N_A is Avogadro's number) we can rewrite Eq. (3) as

$$[n_A]_s \left[\frac{\text{molecules}}{\text{cm}^2}\right] = \frac{\alpha P \, N_A}{\sqrt{2\pi M_A RT}} \, \tau_0 \, e^{\Delta H_{ads}/RT}$$

$$= 3.52 \times 10^{22} \frac{P_{torr}}{\sqrt{M_A T}} \, \tau_0 \, e^{\Delta H_{ads}/RT}. \tag{4}$$

From the knowledge of P, T and ΔH_{ads}, $[n_A]_s$ can be estimated. For example, assuming that $\tau_0 = 10^{-12}$ s and $\alpha = 1$, $\Delta H_{ads} = 2$ kcal/mole and T = 300 K, the surface concentration of argon at P = 10^{-6} torr is immeasurable, $[n_A]_s \approx 10^4$ molecules/cm^2 (one monolayer is about 10^{15} molecules/cm^2). It is still a fraction of a monolayer at one atmosphere. However, at T = 100 K the surface is saturated with a monolayer of argon at one atmosphere ($\sim 10^{15}$ molecules/cm^2). For a higher value of ΔH_{ads}, say 15 kcal/mole, the surface is covered with a measurable quantity (1–100% of a monolayer) of gas at 300 K even at 10^{-6} torr. Gas-surface systems that are characterized by weak interactions ($\Delta H_{ads} < 15$ kcal/mole accompanied by short residence times), that require adsorption studies to be carried out at low T and at relatively high pressure ($\sim$1 atm), are called physical adsorption systems. Adsorbates that are characterized by stronger chemical interactions ($\Delta H \geqslant 15$ kcal/mole) where near-monolayer adsorption commences even at 300 K and at low pressures, $\leqslant 10^{-6}$ torr, are called chemisorbed systems. Although these traditional names imply two distinct types of adsorption the various gas-surface systems exhibit a gradual change from the physisorption to the chemisorption regime.

The coverage, θ, may be varied by changing the pressure over the surface while maintaining a well-chosen constant temperature. The θ vs P(T) curve so obtained for any given gas-surface system is called the adsorption isotherm. The simplest adsorption isotherm is obtained from Eq. (4) which we can rewrite as

$$\theta = k''P \tag{5}$$

with

$$k'' = \frac{1}{n_0} \frac{\alpha N_A}{\sqrt{2\pi M_A RT}} \tau_0 \, e^{\Delta H_{ads}/RT} .$$

Thus the coverage is proportional to the first power of the pressure at a given temperature provided that we have an unlimited number of adsorption sites on the surface and that ΔH_{ads}, which reflects the nature of the gas-surface interaction, does not change as the coverage is changing. Langmuir[1)] has derived a different adsorption isotherm that has become very useful in describing many adsorption processes that terminate when a monolayer coverage is reached. He assumed that any gas molecule that strikes an adsorbed molecule must reflect from the surface while it adsorbs when impinging on the bare surface. If $[n_0]$ is the surface concentration on a completely covered surface, the concentration of surface sites available for adsorption after building up an adsorbate concentration $[n_A]_s$ is $[n_0] - [n_A]_s$. Of the flux, F, striking the surface a fraction $([n_A]_s/[n_0])$ will strike molecules already adsorbed and therefore be reflected. Thus a fraction $(1 - [n_A]_s/[n_0])$ of the total incident flux will be available for adsorption. As a result Eq. (3) is modified to

$$[n_A]_s = \left[1 - \frac{[n_A]_s}{[n_0]}\right] F\tau \tag{6}$$

which can be rearranged to yield the Langmuir adsorption isotherm:

$$\theta = \frac{k''P}{1 + k''P} . \tag{7}$$

Typical adsorption isotherms that obey Eqs. (5) and (7) are shown in Figs. 2.1 and 2.2, respectively. It should be noted that a linear Langmuir plot can be obtained by plotting $1/[n_A]_s$ against $1/P$ where the slope is $1/k$ and the intercept is $1/[n_0]$ as seen after rearrangement of Eq. (7). The adsorption isotherms are utilized primarily to determine the surface area of porous solids and the heats of adsorption. The isotherms yield the amount of gas adsorbed. By multiplying with the area occupied per molecule

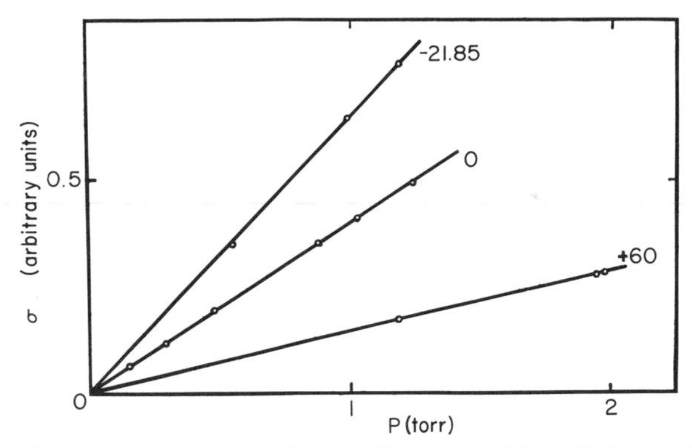

Fig. 2.1. Adsorption isotherms of argon on silica gels (σ stands for θ; labels on curves indicate different temperatures in °C)

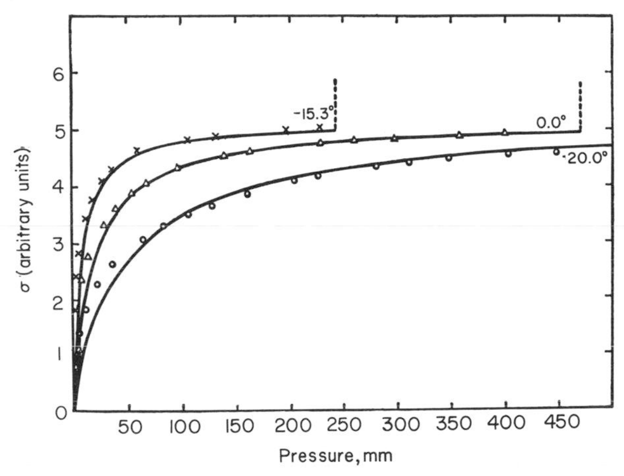

Fig. 2.2. Adsorption isotherms of ethyl chloride on charcoal (σ stands for θ; labels on curves indicate different temperatures in °C)

that is determined independently, the total surface area is determined. For example, the area per molecule is 16.2 Å^2 for N_2 and 25.6 Å^2 for krypton on a large variety of surfaces. The heat of adsorption is obtained from adsorption isotherms measured at different temperatures using the Clausius-Clapeyron equation

$$\left[\frac{d \ln P}{d(1/T)}\right]_{\theta\,=\,\text{const.}} = -\frac{\Delta H_{ads}}{R} \quad . \tag{8}$$

Reference

1. Langmuir, I: J. Am. Chem. Soc. *40*, 1361 (1918)

III Principles of Ordering of Adsorbed Monolayers

1 Causes of Ordering

Once a molecule lands on the solid surface it may slide along the surface plane or remain bound at a specific site during much of its surface residence time. As long as ΔH_{ads} and the activation energy for bulk diffusion $\Delta E^*_{D(bulk)}$ are high enough as compared to kT ($\geqslant$10 kT), we are assured of a residence time that is long enough to permit thermal equilibration among the adsorbates and between the adsorbate and substrate atoms, i.e., adsorption. Ordering, however, primarily depends on the depth of the potential energy barrier that keeps an atom or molecule from hopping to a neighboring site along the surface. The activation energy for surface diffusion, $\Delta E^*_{D(surface)}$ is an experimental parameter that is of the magnitude of this potential energy barrier. $\Delta E^*_{D(surface)}$ may be obtained for self-diffusion or for the diffusion of adsorbates on well-characterized surfaces by several techniques. Among them field ion microscopy[1)] and sinusoidal wave analysis[2)] are the most prominent at present. The $\Delta E^*_{D(surface)}$ for Ar, W adatoms and of O atoms on tungsten surfaces are 2 kcal, 15 kcal and 10 kcal, respectively. For small values of $\Delta E^*_{D(surface)}$ ordering is restricted to low temperatures since the adsorbate atoms become very mobile as the temperature is increased. For higher values of $\Delta E^*_{D(surface)}$ ordering cannot commence at low temperatures since the adsorbate atoms need to have a considerable mean free path along the surface to find their equilibrium position once they landed on the surface at a different location. Of course if the temperature is too high the adsorbed atoms or molecules desorb or vaporize.

It should be noted that in the limit of very large heats of chemisorption one may form surface compounds, oxides or carbides, for example. In this circumstance ordering of the new surface phase may require the relocation of the substrate atoms as well as the adsorbate atoms. Such chemisorption-induced reconstructions have been observed for several systems and its presence makes the conditions necessary for ordering in the surface layer very difficult to analyze indeed. Some of these systems will be discussed later in this paper.

The interatomic forces responsible for the binding of adsorbates at surfaces and for the ordering of overlayers are of various types. The binding of adsorbates to substrates is frequently due to the strong covalent chemical forces, as a result of the presence of electron orbitals overlapping both the substrate and the adsorbate. Some adatoms (notably the rare gases) and many molecules will only weakly stick to substrates.

The binding force is then predominantly due to the Van der Waals interaction and we have physisorption.

The binding forces have components perpendicular and parallel to the surface. The perpendicular component is mainly responsible for the binding energy (heat of adsorption), while the parallel component often determines the binding site along the surface. The binding site may however also be affected by adsorbate-adsorbate interactions, which are also responsible for any ordering within an overlayer. These interactions may be arbitrarily subdivided into direct adsorbate-adsorbate interactions (not involving the substrate at all) and substrate-mediated interactions: the latter are complicated many-atom interactions. Dipole-dipole interactions are an example of such interactions, involving the exact charge distribution of the adsorbed particles, the shape of the electrostatic dipolar fields at the surface and, of course, self-consistency requirements, since dipolar charge distributions are themselves affected by nearby dipoles.

The adsorbate-adsorbate interactions can be repulsive; they always are repulsive at sufficiently small adsorbate-adsorbate separations. They may be attractive at larger separations, giving rise to the possibility of island formation. They may be oscillatory, changing back and forth between attractive and repulsive as a function of adsorbate-adsorbate separation, with a period of several angstroms, giving rise, for example, to non-close-packed islands[3]. Such is the case of oxygen adsorbed on W(110), for instance[4]. And they usually are anisotropic, differing according to the orientation of the lines connecting pairs of adsorbates, since the single-crystal substrate surface is inherently anisotropic. Additional anisotropy occurs with many-adsorbate interactions [as observed for oxygen on W(110)[4]], as one can easily illustrate for a single adsorbate near a cluster of two adsorbates: it may be favorable to produce a 3-in-line cluster, or instead an L-shaped cluster. This particular form of interaction is ideally studied with Field Ion Microscopy, especially by observing the diffusion of such clusters along surfaces[5]. However, the analysis of such observations is only in its early stages.

Except for the strong repulsion at close separations which prevents adsorbates from penetrating each other, the adsorbate-adsorbate interactions are usually weak as compared to the adsorbate-substrate interactions, even when one only considers the components of the forces parallel to the surface. Thus, in the case of chemisorption (which occurs primarily with single-atom adsorption), where the adsorbate-substrate interaction dominates, one finds that the adsorbates usually choose an adsorption site that is independent of the coverage and of the overlayer arrangement, i.e., independent of which positions the other adsorbates choose. As will be discussed in more detail in Sect. VI, this adsorption site is usually the site that provides the largest number of nearest substrate neighbors, which is indeed independent of the position of other adsorbates. Adsorbates with these properties normally do not accept close-packing: the substrate controls the overlayer geometry and imposes a unique adsorption site. Close-packing of an adsorbate layer is, however, often observed with other adsorbates. Then the overlayer chooses its own lattice (normally a hexagonal close-packed arrangement) with its own lattice constant, independently of the substrate lattice: so-called incommensurate lattices form. In this case no unique adsorption site exists: each adsorbate is differently situated with respect to the substrate. This situation is especially common in the physisorption of rare gases with its relatively weak adsorbate-substrate interactions, which therefore allows the adsorbate-adsorbate interactions to play the dominant role in determining the overlayer geometry. Sometimes the substrate imposes a particular orientation on the overlayer lattice in this circumstance.

The chemisorption case is exemplified by oxygen and sulfur on metals, the physisorption case by krypton and xenon on metals and graphite. Intermediate cases do exist: for example, undissociated CO on metals is not physisorbed but chemisorbed and nevertheless it seems in many cases to be able to produce close-packed hexagonal overlayers. Also, some metal surfaces [for example, Pt(100), Ir(100), Au(100)] reconstruct into different lattices, exhibiting the effect of adsorbate-adsorbate interactions (here the adsorbate is just another metal atom of the same species as in the substrate).

As will be seen in Sect. V, the variety of possible ordered surface structures is immense. This is a reflection of the large number of possible relative magnitudes of the various forces responsible for the bonding and the ordering. When one realizes that each of these forces varies in three dimensions, often drastically, it is not surprising that a very large number of combinations and therefore of structures is possible. It may be true that every conceivable two-dimensional ordering arrangement is possible in nature on surfaces, even with simple adsorbates on simple surfaces.

The theory of the binding of single adsorbates to substrates is today understood to some extent (cf. Sect. IV and VI), while the theory of adsorbate-adsorbate interactions and especially of large-scale ordering is in its infancy.

2 The Degree of Ordering

A perfectly-ordered surface represents the energetically most favorable surface configuration. However, no real surface is perfectly ordered. There are several reasons for this. Firstly, there is always some thermal energy available to make an adsorbate jump into an energetically less favorable configuration: for example, adsorbate atoms in an ordered overlayer can jump out of registry. Even at zero temperature, the zero-point motion gives rise to disorder in the form of vibrations about the atomic equilibrium positions. Secondly, in no experiment is the surface allowed to reach the asymptotic equilibrium: some forms of disorder have characteristic half-lives of the order of many hours. Thus an adatom trapped interstitially in a normally unoccupied site of a c(2×2) overlayer arrangement on a square-lattice substrate, will, at low enough temperatures, have very little chance of migrating to a proper unoccupied site prescribed by the c(2×2) lattice, since such a site may be located at a considerable distance. Other examples of long half-life disorder are steps in the surface (if they are undesirable), bulk defects extending to the substrate surface and, of course, impurities.

Some forms of disorder common in adsorbed layers are: islands of clustered adsorbates leaving patches of bare substrate; domains in which different patches of the overlayer have identical structure but do not match at their junction because of an error in registry, i.e., an error in relative positioning parallel to the surface; periodicity errors, in which individual adsorbates do not fit in the periodic arrangement of the surrounding adsorbates; these periodicity errors subdivide into those that involve inequivalent sites (e.g., an adsorbate choosing a two-fold bridge site, while the overlayer as a whole involves only four-fold hollow sites) and those that retain unique adsorption sites, but the improper ones; there are the cases of individual disorder (one adsorbate in a wrong position) and the cases of collective disorder (such as phonons and liquid layers).

While perfect order is never present, perfect disorder also does not exist at surfaces. In the liquid (or the gaseous) state of overlayers on surfaces, the adsorbates cannot pass through each other: this gives rise to a limited amount of short-range order. Additionally, there is always some non-zero parallel component of the substrate-adsorbate interaction that will make the adsorbates spend more of their time at one type of location than at others: this also is a form of ordering.

Surface-sensitive diffraction techniques, especially LEED, can in principle detect any kind of ordering or disordering at a surface. The exact state of a surface at any moment can be represented by a Fourier series that describes the surface in terms of all possible periodicities (Fourier components). Each different periodicity present in an overlayer produces diffraction into a well-defined direction specified by the period and the orientation of the particular periodicity. The intensity of the diffraction measures the amount of order with that periodicity (this intensity is modulated by the surface structure perpendicular to the surface and can therefore not be taken to be a direct measure of the amount of order without proper care). Therefore, diffraction methods allow one to easily filter out many forms of disorder: in LEED one may analyze just the sharp spots observed on a screen and thereby filter out all disorder that has periodicities defined by points between those spots. Non-diffraction methods do not have this kind of disorder-filtering capability: they usually average over all information, whether from the ordered part of the surface or the disordered part of the surface. Non-diffraction methods may sometimes have other types of disorder filtering, however: if disorder produces features at different energies of a spectrum, for example, such energies might be screened out. Thus in High Resolution Electron Energy Loss Spectroscopy, undesired surface adsorbates could produce resonance levels at different energy losses, which can then be ignored.

As an example of the analysis of the ordering of an overlayer of adsorbates, we may take the question of detecting island formation. LEED provides a means for identifying when island formation takes place[3,4] although it does not always give a definitive answer. To monitor island formation the presence of adsorbate-induced extra spots in the diffraction is necessary. Thus the adsorbate must produce a superlattice and we assume this case in the following discussion.

To recognize island formation one takes advantage of the difference between coherent and incoherent diffraction from a set of N identical scatterers. If the waves scattered off the individual scatterers are incoherent in their phases, the observed intensity will be proportional to N (addition of intensities). If, however, these scattered waves are coherent, the intensity will be proportional to N^2 (addition of amplitudes). Incoherence occurs either when the incident wave arrives with incoherent phases at different scatterers, which occurs in practice for scatterers separated by at least the "coherence length" of the incident beam, or when the scatterers themselves are located incoherently, i.e. are disordered.

The key to the detection of island formation is the coherence length (also called instrument response width or transfer width) of the incident electron beam, typically 100 Å. If the coherence length were much smaller or much larger than 100 Å, one would not obtain information about island formation from LEED. If the coherence length were variable, this degree of freedom would be valuable to study island formation on different scales (to a limited extent, it is variable, namely by changing the angle of incidence or the electron energy, but more flexibility would be useful). At low coverages, if islands form that are smaller than the coherence length and also farther apart than the coherence length, then a diffraction pattern characteristic of an island is produced: each of the extra spots has a sharpness inversely proportional to the island diameter. The extra-spot intensity is then proportional to the square of the coverage, if one assumes that additional adsorbates will attach themselves to islands. If instead they initiate new islands (still far apart), the intensity would increase linearly with coverage; with such island birth, the spot sharpness is constant. In reality both island growth and island birth can take place simultaneously in varying proportions, depending on such factors as the surface mobility of the adsorbate and the binding energy of adsorbates to islands: then the extra-spot intensities would vary with a law between the first and second power of the coverage. The extra-spot sharpness would simultaneously be less than inversely proportional to the island diameter, i.e., more constant. In contrast, non-island adsorption at low coverages gives no extra spots, but a weak diffuse background.

However, as soon as the coverage becomes sufficiently high that either the island diameter is at least equal to the coherence length or the island-island distance is at most equal to the coherence length, these relations change. With an island diameter at least equal to the coherence length, the extra-spot sharpness saturates at a value determined by the coherence length, while the extra-spot intensities become linear with coverage. With an island-island distance at most equal to the coherence length, the extra spots remain relatively diffuse with increasing coverage (due to a relatively constant and small island size), while these spots weaken again due to antiphase domains (in some cases extra-spot splitting occurs). For comparison, non-island adsorption at these higher coverages produces either no order at all (no extra spots) or else spots with a sharpness determined by the ordering distance and with an intensity quadratic in the coverage.

Complications in actual studies along the lines described above come from uncertainties in the question of island growth vs. island birth and a lack of understanding of the factors determining these. Also the range of the ordering forces plays a role that should be explored more systematically than hitherto. And, of course, the bonding configuration may change with coverage, causing a change in intensities not related to the effects described above.

Generally, it is difficult to obtain experimental information about the exact form of disorder present on any actual surface: much work remains to be done in this direction.

3 The Effect of Temperature on the Ordering of Adsorbed Monolayers

In Fig. 3.1 the influence of temperature on the ordering of C_3–C_8 saturated hydrocarbon molecules on the Pt(111) crystal face is shown. At the highest temperatures adsorption may not take place, since under the exposure conditions the rate of desorption is greater than the rate of condensation of the vapor molecules. As the temperature is decreased the surface converage increases and ordering becomes possible. First, one-dimensional lines of molecules form, then upon further dropping of the temperature ordered two-dimensional surface structures form. Not surprisingly, the temperatures at which these ordering transitions occur depend on the molecular weights of the hydrocarbons which also control their vapor pressure, their heats of adsorption and their activation energies for surface diffusion. As the temperature is further decreased multilayer adsorption may occur and epitaxial growth of crystalline thin films of hydrocarbon commences.

Figure 3.1 demonstrates clearly the controlling effect of temperature on the ordering and the nature of ordering of the adsorbed monolayer. Although changing the pressure at a given temperature may be used to vary the coverage by small amounts and thereby change the surface structures in some cases, the variation of temperature has a much more drastic effect on ordering. All of the important ordering parameters (the rates of desorption, surface and bulk diffusion) are exponential functions of the temperature.

An example of the control of surface diffusion on ordering is shown in Fig. 3.2. Naphthalene forms a poorly ordered structure when adsorbed on a Pt(111) crystal face

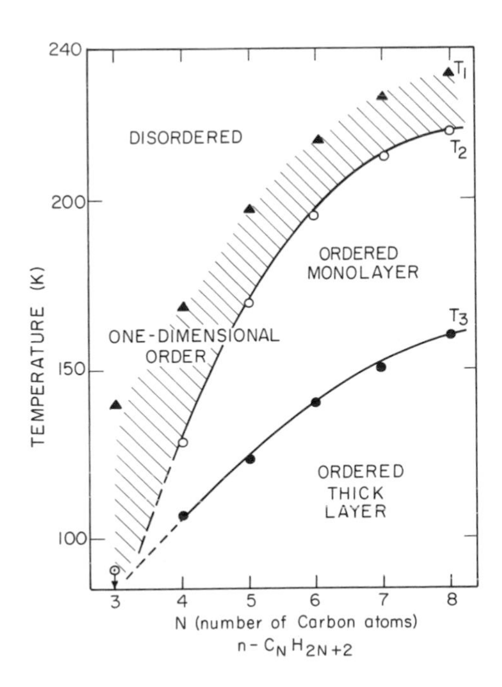

Fig. 3.1. Monolayer and multilayer surface phases of the n-paraffins C_3–C_8 on Pt(111), and the temperatures at which they are observed at 10^{-7} Torr

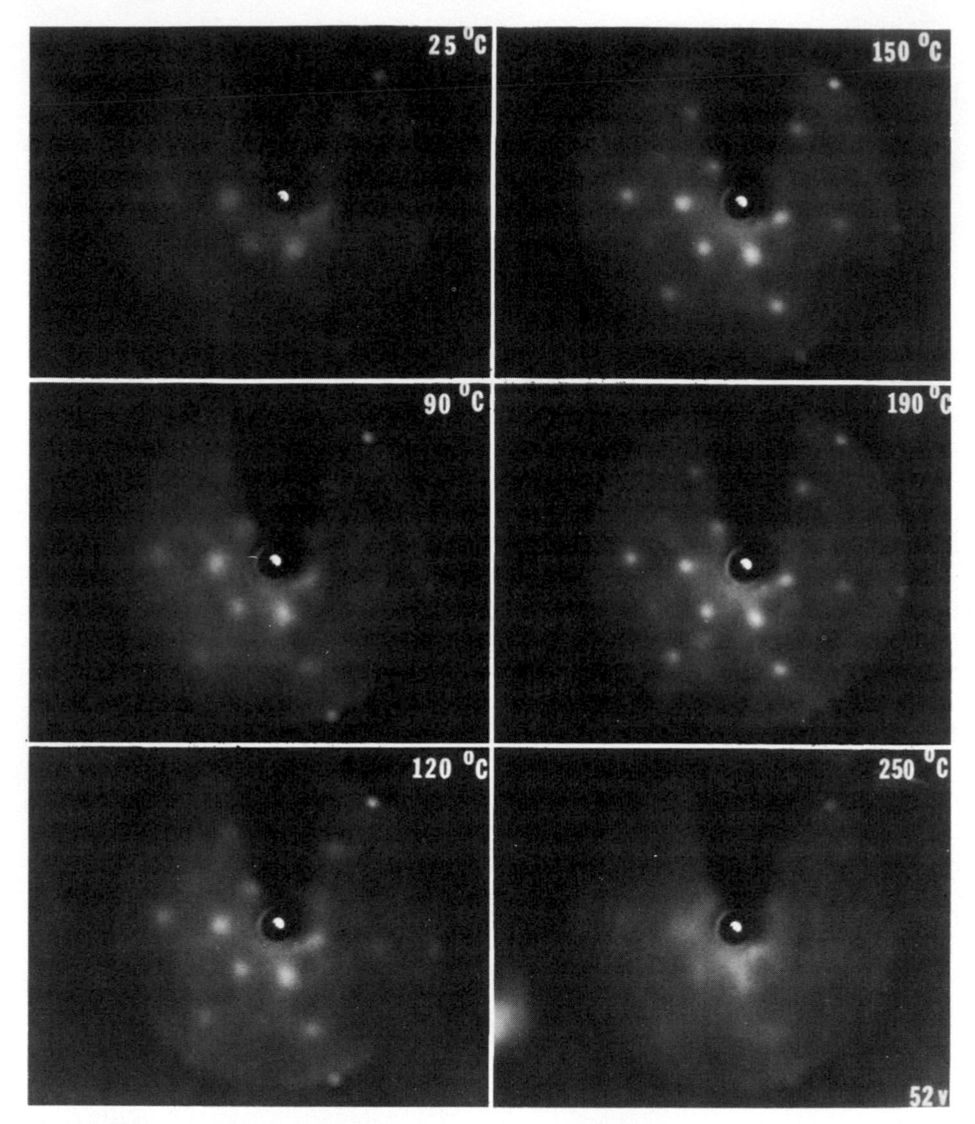

Fig. 3.2. Electron diffraction pattern from a monolayer of naphthalene on Pt(111) for an electron energy of 52 eV as a function of temperature. Sharp spots correspond to good ordering

at 300 K. Upon heating the almost glassy layer to 450 K a well ordered (6×6) surface structure forms. For large molecules surface diffusion plays a visibly important role in ordering as detected by several investigations.

Temperature also markedly influences the chemical bonding to the surfaces. There are adsorption states that can only be populated if the molecule overcomes a small potential energy barrier. The various bond breaking processes are similarly activated. The adsorption of most reactive molecules, on chemically active solid surfaces, takes place without bond breaking at sufficiently low temperatures. As the temperature is increased bond breaking occurs sequentially until the molecule is atomized. Thus the chemical nature of the molecular fragments will be different at various temperatures. There is almost always a temperature range, however, for the ordering of intact molecules in chemically active adsorbate-substrate systems. It appears that for these systems ordering is restricted to low temperatures below 150 K and consideration of surface mobility becomes perhaps secondary.

4 The Effect of Surface Irregularities on Ordering

When a solid surface is viewed under the optical or the electron microscope it almost always exhibits a large degree of roughness on the macroscopic scale. There are protruding hills of several hundred atomic layers in height and discontinuities that separate relatively smooth terraces. A typical electron microscope picture of an etched platinum single crystal surface is shown in Fig. 3.3. On the atomic scale, however, the surface appears to be much smoother. The very high quality low energy electron diffraction pattern commonly observed from most cleaned and annealed solid surfaces must require the presence of domains of ordered atoms of larger than 100–200 Å in diameter. The coherent scattering of electrons that yield the sharp, small and high intensity diffraction spots can only occur if the size of the scattering areas is larger than the electron coherence length. Were the ordered domains smaller a broadening of the diffraction spots would be observed which is in fact what happens if the surface is roughened by ion bombardment, for example. Another technique, field ion microscopy (FIM), which can display the surface topography of a small tip of $\sim 10^{-4}$ cm diameter with atomic resolution, also shows the large degree of atomic order that is possible at surfaces.

As long as nucleation is an important part of the adsorbate ordering process, surface roughness is likely to play an important role in preparing ordered surface structures. It is observed frequently that the ease of ordering and the quality of the ordered surface structures of adsorbate changes from one substrate sample to another. There is often great "improvement" in the ordering characteristics right after ion bombardment cleaning and brief thermal annealing of the substrate surface, then ordering becomes better as the substrate is annealed and thereby ordered more and more. The transformation temperature or pressure at which one adsorbate surface structure converts into another can also be affected by the presence of uncontrolled surface irregularities. Although the surface structures of adsorbates by and large are reproducible from sample to sample and laboratory to laboratory, the uncertainties in the experimental conditions necessary to form the ordered surface structures are caused most frequently by uncontrolled surface defects. The other causes that could influence ordering are the presence of small amounts of surface impurities that block nucleation sites or interfere with the kinetics of ordering or impurities below the surface that are pulled to the surface during absorption and ordering.

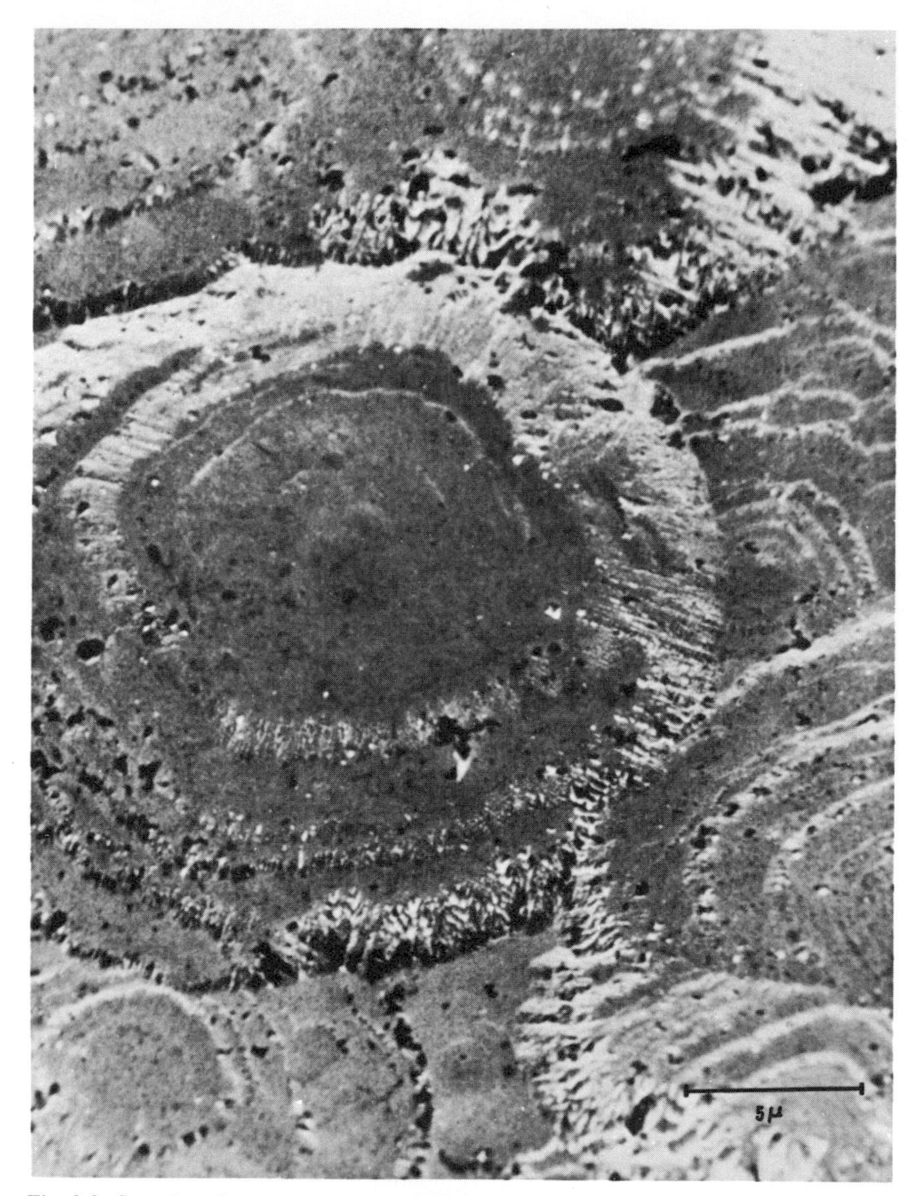

Fig. 3.3. Scanning electron microscope (SEM) photograph of an etched Pt(100) surface tilted at 45° to the incident electron beam to enhance picture contrast

It is much easier to investigate the effect of surface irregularities on ordering using stepped crystal surfaces. Unlike in the case of uncontrolled surface defects on a (111) face of an fcc metal for example, steps are readily detectable by LEED or FIM. They are likely to be ordered with a well-defined periodicity and surfaces can be prepared in such a way that steps are the predominant highest concentration surface defects.

The influence of atomic height steps on the ordering of adsorbate structures has been investigated in several studies[6]. In general, the smaller is the ordered terrace between steps the stronger is the effect of steps on ordering. The ordering of small molecular adsorbates on a 6(111) × (100) rhodium surface was largely unaffected by the presence of steps[6a]. However, ordering was influenced by steps on the larger step-density Rh(331) crystal face[6a]. Nitrogen and carbon layers were observed to extend over several terraces on stepped copper surfaces[6b] and the ordering of Ar and Kr was unaffected by the presence of steps on copper and silver (211) crystal faces[6c]. Just as in the case of uncontrolled irregularities, steps can markedly affect the nucleation of ordered domains. It is frequently observed on W and Pt stepped surfaces that when 2 or 3 equivalent ordered domains may form in the absence of steps, only one of the ordered domains grows in the presence of steps. Oxygen surface structures exhibit this phenomenon as well and have been studied in the greatest detail.

In many cases ordering is no longer observable in the presence of steps. Ordered carbonaceous layers form on the Ir(111) crystal face, for example, while ordering is absent on the stepped iridium surface. Ordering is absent on stepped Pt surfaces for most molecules that would order on the low Miller-Index (111) or (100) surfaces.

In some cases, the step sites have different chemistry, i.e., they break chemical bonds, thereby producing new chemical species on the surface. This happens for example during NO adsorption on a stepped platinum surface[7]. In this circumstance the step effect on ordering is through the new types of chemistry introduced by the presence of steps. Hydrocarbons for example dissociate readily at stepped surfaces of platinum or nickel while this occurs much more slowly on the low Miller-Index surfaces in the absence of a large concentration of steps[8]. As a result ordered hydrocarbon surface structures cannot be formed on the stepped surfaces of these metals while they can be produced on the low Miller-Index surfaces.

There is also a great deal of evidence for increased sticking probability at stepped surfaces. The change in the magnitude of the adsorption probability ranges from 20% to orders of magnitude. Also, several studies revealed increased binding energies at step sites. 10–20% increase in binding energies at steps on Ni and Pt surfaces are common. Both the increased adsorption probability and binding energies at steps may strongly affect the kinetics of ordering. Thus, there are many reasons for the different ordering characteristics of adsorbed monolayers in the presence of surface irregularities.

5 Unit Cell Notations

In the majority of cases where adsorbates form ordered surface structures, the unit cells of those structures are larger than the unit cell of the substrate: the surface lattice is then called a super lattice. The surface unit cell is the basic quantity in the description of the ordering of surfaces. It is necessary therefore to have a notation that allows the unique characterization of superlattices relative to the substrate lattice.

Two common notations are used to relate superlattices to substrate lattices, one of these notations being a simplification of the other for simple cases. Let the substrate surface lattice be characterized by a pair of basis vectors $(\vec{a}_1, \vec{a}_2)$ spanning the unit cell, and similarly the superlattice is represented by $(\vec{b}_1, \vec{b}_2)$. These pairs of vectors are in general related by a matrix M:

$$\begin{pmatrix} \vec{b}_1 \\ \vec{b}_2 \end{pmatrix} = \begin{pmatrix} M_{11} & M_{12} \\ M_{21} & M_{22} \end{pmatrix} \begin{pmatrix} \vec{a}_1 \\ \vec{a}_2 \end{pmatrix} = M \begin{pmatrix} \vec{a}_1 \\ \vec{a}_2 \end{pmatrix} .$$

This is the matrix notation, in which the matrix M uniquely characterizes the relationship between the unit cells (note that the concept of unit cell is not unique – different unit cells can describe the same lattice – and so different matrices M can characterize the relationship between two given lattices).

A non-matrix notation, called Wood notation[9], can be used when the angles between the pairs of basis vectors are the same for the substrate and the superlattice, i.e., when the angle between $\vec{a}_1$ and $\vec{a}_2$ is the same as the angle between $\vec{b}_1$ and $\vec{b}_2$. Then the unit cell relationship is given by, in general,

$$\text{c or p } (v \times w) \, R\alpha .$$

Here v and w are the elongation factors of the basis vectors, i.e.,

$$v = |\vec{b}_1| / |\vec{a}_1|, \; w = |\vec{b}_2| / |\vec{a}_2| ,$$

while α is the angle of rotation between the lattices, i.e., the angle between $\vec{a}_1$ and $\vec{b}_1$. The prefixes "c" and "p" mean "centered" and "primitive", respectively, with "centered" representing the case where an adsorbate is added in the center of the primitive $(v \times w) \, R\alpha$ unit cell. The prefix p is optional and often omitted, while the suffix $R\alpha$ is omitted when $\alpha = 0$.

For illustration, we list in Table 3.1 a number of superlattices found commonly in overlayers.

We should mention here a special notation used for describing high-Miller-index surfaces. Such surfaces can often be more usefully described as stepped surfaces involving relatively close-packed terraces of low-Miller-index orientation separated by steps whose faces have also a low-Miller-index orientation. For example, the fcc(755) surface can be more easily visualized with the notation fcc(S)-[6(111) × (100)], where (S) means "stepped", since this indicates that the surface is composed of terraces of (111) orientation and 6 atoms wide, separated by steps of (100) orientation and 1 atom high. A list of such correspondences of notation for stepped fcc surfaces is included in Sect. V.

6 Unit Cells

The unit cells of adsorbate layers are primarily a function of the coverage: as the coverage varies, many adsorbates produce complete series of successive different unit

Table 3.1. Wood and matrix notation for a variety of superlattices on low Miller index crystal surfaces

Substrate	Overlayer unit cell	
	Wood notation	Matrix notation
fcc(100), bcc(100)	p(1×1)	$\begin{pmatrix}1 & 0\\0 & 1\end{pmatrix}$
	c(2×2) = ($\sqrt{2}$× $\sqrt{2}$)R45°	$\begin{pmatrix}1 & -1\\1 & 1\end{pmatrix}$
	p(2×1)	$\begin{pmatrix}2 & 0\\0 & 1\end{pmatrix}$
	p(1×2)	$\begin{pmatrix}1 & 0\\0 & 2\end{pmatrix}$
	p(2×2)	$\begin{pmatrix}2 & 0\\0 & 2\end{pmatrix}$
	(2$\sqrt{2}$× $\sqrt{2}$)R45°	$\begin{pmatrix}2 & 2\\-1 & 1\end{pmatrix}$
fcc(111) (60° between basis vectors)	p(2×1)	$\begin{pmatrix}2 & 0\\0 & 1\end{pmatrix}$
	p(2×2)	$\begin{pmatrix}2 & 0\\0 & 2\end{pmatrix}$
	($\sqrt{3}$× $\sqrt{3}$)R30°	$\begin{pmatrix}1 & 1\\-1 & 2\end{pmatrix}$
fcc(110)	p(2×1)	$\begin{pmatrix}2 & 0\\0 & 1\end{pmatrix}$
	p(3×1)	$\begin{pmatrix}3 & 0\\0 & 1\end{pmatrix}$
	c(2×2)	$\begin{pmatrix}1 & -1\\1 & 1\end{pmatrix}$
bcc(110)	p(2×1)	$\begin{pmatrix}2 & 0\\0 & 1\end{pmatrix}$

cells, cf. Sect. V. The coverage is defined here in such a way that unit coverage, $\theta = 1$, occurs when the adsorbate occupies all equivalent adsorption sites.

One can correlate the coverage θ with certain features of the unit cells that the adsorbates can adopt on surfaces. Let us define S to be the area of the substrate unit cell.

When $1/\theta$ is an integer n, there are n substrate unit cells per adsorbate, and a superlattice with unit cell area nS can occur. Thus for $\theta = 1/2$, a superlattice with unit cell area 2S may exist, examples of which are designated p(2×1) and c(2×2), and illustrated in Fig. 3.4.

When $1/\theta$ is a rational number m/n (m,n integer, undivisible), there are m substrate unit cells per set of n adsorbates: a superlattice with unit cell area mS can form, with the superlattice unit cell containing n arbitrarily-positioned adsorbates. It may happen in this case that the adsorbates between themselves (ignoring the substrate) form a structure that has a smaller unit cell than the superlattice unit cell: one must then distinguish between the overlayer unit cell (defined in the absence of a substrate) and

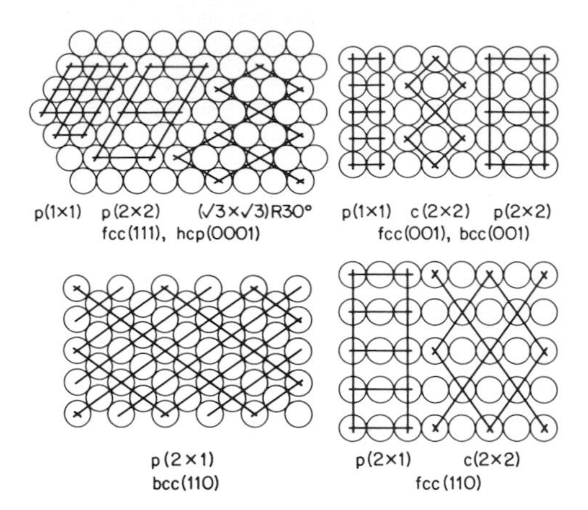

Fig. 3.4. Common superlattices on low Miller index crystal surfaces. The Wood notation is used

the so-called coincidence unit cell (describing the combined substrate-overlayer system). An example is shown in Fig. 3.5 for the case of Pd(100) + $(2\sqrt{2} \times \sqrt{2})$R45° 2CO, which has two molecules per coincidence unit cell[10]. Note that this adsorbate has managed to combine bridge sites with an approximately hexagonal arrangement.

When $1/\theta$ is an irrational number, the overlayer lattice bears in general no relationship to the substrate lattice: the surface unit cell becomes infinite and the unit cell areas become incommensurate. This case corresponds to totally independent lattices, as is approximated by physisorbed systems.

In practice the distinction between rational and irrational values of $1/\theta$ is unimportant, because LEED cannot distinguish between unit cells larger than the coherence distance of the electron beam (~100 Å). It is customary to designate as incommensurate any overlayer that produces a coincidence unit cell larger than the LEED coherence distance. In fact, a truly incommensurate overlayer is impossible, since it could only occur in the limit of vanishing adsorbate-substrate forces parallel to the surface.

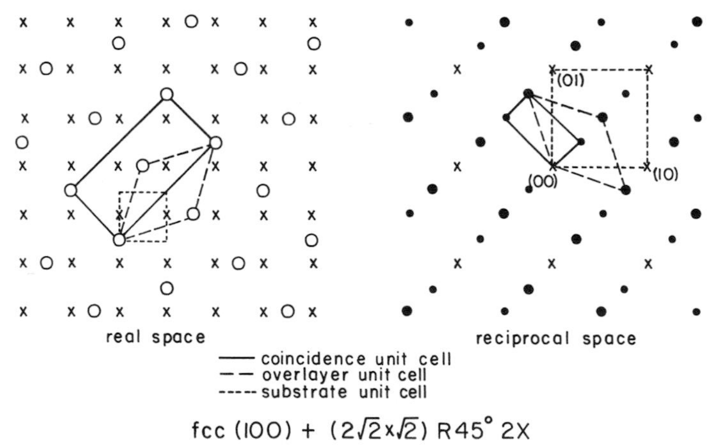

Fig. 3.5. Real space (*left*) and reciprocal space (*right*) applicable to Pd(100) + $(2\sqrt{2} \times \sqrt{2})$ R45° 2CO. Substrate (*overlayer*) atoms and diffraction spots are indicated by crosses (*circles*). Large filled circles represent kinematically produced spots, small filled circles represent multiple-diffraction spots

References

1a. Müller, E. W., Tsong, T.T.: Field Ion Microscopy, New York: Elsevier 1969
b. Müller, E. W.: Science *149*, 591 (1965)
2. Bonzel, H. P., Gjostein, N. A.: App. Phys. Lett. *10*, 258 (1967)
3. Somorjai, G. A., Farrell, H. H.: Adv. Chem. Phys. *20*, 215 (1971)
4. Wang, G.-C., Lu, T.-M., Lagally, M. G.: J. Chem. Phys. *69*, 479 (1978)
5. Ehrlich, G.: Surf. Sci. *63*, 422 (1977)
6a. Castner, D. G., Somorjai, G. A.: Surf. Sci. *83*, 60 (1979)
b. Perdereau, J., Rhead, G. E.: Surf. Sci. *24*, 555 (1971)
c. Roberts, R. H., Pritchard, J.: Surf. Sci. *54*, 687 (1976)
7. Gland, J. L.: Surf. Sci. *71*, 327 (1978)
8a. Lang, B., Joyner, R. W., Somorjai, G. A.: Surf. Sci. *30*, 454 (1972)
b. Erley, W., Wagner, H., Ibach, H.: to be published
9. Wood, E. A.: J. Appl. Phys. *35*, 1306 (1964)
10. Park, R. L., Madden, H. H.: Surf. Sci. *11*, 158 (1968)

IV Methods of Structure Analysis

1 Introduction

In surface structure determinations the basic information sought is the relative atomic positions at crystal surfaces on a scale of between roughly 0.001 and 100 Å. This covers the knowledge of unit cell shapes, sizes and orientations, of bond lengths, of bonding geometry (bonding site and bond angles), of the internal structure of adsorbed molecules and of the degree of ordering and disordering within surface layers. For an understanding of surface processes, geometrical information is needed for the clean surface (which may be reconstructed or not with respect to the bulk geometry), and for adsorbed atoms and molecules on such clean surfaces, including the penetration of atoms into the surface.

In this chapter we review the various experimental techniques that are used to study the arrangement of atoms and molecules at surfaces. First we discuss the basic aspects of the sample preparation, since this is a particularly critical step in surface studies. Then we discuss the principles involved in the measurements of surface-specific physical quantities. Since each of the many techniques of surface analysis is sensitive to a few particular aspects of the surface (such as relative atomic positions, electronic levels, chemical composition, binding energies and vibration frequencies), we classify these techniques according to the surface characteristic that they are most sensitive to.

2 Sample Preparation

Studies of surfaces all begin with sample preparation. Ideally we would like to use surfaces whose surface composition and atomic structure is uniform. For studies of thermodynamically stable surfaces it is often necessary to heat the specimen to be used for surface studies to an elevated temperature in a controlled, chemically inert ambient (or in vacuum) to achieve equilibrium of the surface and bulk compositions and to remove excess defects that were introduced during previous specimen preparation. Frequently we desire to study ordered single crystal surfaces. In this circumstance single crystal rods or boules that were grown by zone referring, vapor transport or stain annealing are oriented by the back-reflection Laue technique and then cut along the desired crystal face. Diamond blade or spark erosion cutting techniques are used most frequently for this purpose. The cutting treatment damages the near surface regions of the specimen often severely (especially if the material is soft) and renders it amorphous. This damaged layer should be removed by chemical or electrochemical dissolution (etching) that should not affect the surface orientation. The preparation of uniform surfaces also includes repeated polishing by fine mesh alumina or carbide particles followed by repeated etching. The sample so prepared is placed into the reaction chamber using suitable holders that permit heating or cooling and accurate positioning. Usually a thermocouple is attached also for accurate temperature determination.

Frequently the preparation of a specimen requires unique experimental conditions and must be performed inside the experimental chamber. For example, for studies of argon or xenon single crystal surfaces or for preparation of surfaces of solid benzene, methane or other high vapor pressure materials a low temperature environment is needed. For the preparation of a certain phase of solids that undergo phase transformations (iron, cobalt, uranium, etc.) again controlled temperature ranges are required during preparation. In these circumstances the specimen may be prepared by vaporization (or vapor transport) onto a well-ordered substrate (generally an ordered single crystal surface) that is held at the desired temperature. The specimen is grown epitaxially until a multilayer (20 to 2000 Å) deposit is produced inside the experimental chamber.

The surface of the specimen that is prepared outside the reaction chamber is usually covered with a thick layer of carbonaceous deposit by the time that it is placed in the chamber. Alternatively, impurities from the bulk of the sample may be diffused to the surface upon heating and segregate there. The impurities most frequently detectable in surface studies that segregate to the surface are carbon, sulfur, silicon, oxygen and aluminium. Some of these impurities may be removed by chemical treatments, by heating the sample in flowing oxygen or hydrogen, etc., to a pressure and temperature where there is a significant rate of solid-gas reactions that remove the impurity without changing the chemical composition of the sample. For the removal of other impurities, the sample surface is usually bombarded with ions of inert gas (argon, xenon or krypton) that are generated inside the experimental chamber using gas pressures in the range of 10^{-5} Torr. Ion fluxes with energies of 100 to 5000 V and sufficient intensities are utilized to remove 1 to 100 layers/s at the near surface region and remove impurities this way. The structural damage introduced by the high energy ion impact can be annealed by heating the specimen thus allowing the surface atoms to move back into their equilibrium position by surface diffusion. In field ion microscopy small-diameter ($\sim 10^{-4}$ cm) crystalline tips are utilized. X-ray absorption fine structure studies permit the use of high surface area porous samples as well as crystalline surfaces. Since heterogeneous metal catalysts are frequently deposited on high surface areas oxides, this technique can be employed for the studies of supported catalysts as well.

Generation of Ultrahigh Vacuum (UHV) and Controlled Processes for Surface Studies. There are two main reasons why high vacuum (10^{-9} to 10^{-4} Torr range) must be maintained around the samples during some phase of the surface chemical experiment. First, it is often desirable to start our investigation with initially clean surfaces: and ultrahigh vacuum (less than 10^{-8} Torr) is needed to achieve a surface that is free from adsorbed gases. Second, many of the surface characterization techniques use electrons or ions as probing particles to reveal the surface structure, composition and oxidation state. These particles need a long mean free path (larger than 10 cm) to be able to strike or exit the sample and then reach the detector without colliding with gas phase molecules. For this reason, pressures lower than 10^{-3} Torr must be used. High vacuum may be generated with many different pumping devices (oil diffusion pump, vacuum ionization pump, sublimation pump, turbo-molecular pump, etc.) and may be maintained indefinitely in a leak-free chamber usually built out of nonmagnetic stainless steel. Vacuum technology has reached a level of sophistication where obtaining high vacuum in short time (less than an hour) has become a simple and reliable procedure.

Often we need to place a specimen in a high pressure environment after surface characterization in high vacuum to carry out our surface studies. The sample may be enclosed by a small high pressure cell which is operated by hydraulic pressure or a threaded drive mechanism. It should be pointed out that the same apparatus can also be used for studies of reactions and processes at the solid-liquid interface. The liquid could readily be introduced and then pumped out after the study and the surface can be studied by the various surface diagnostic techniques before and after the experiments. Using the same principle isolation cells that are capable of containing high pressures around the sample inside the uhv chamber can be constructed in a variety of geometries.

3 Principles of Surface Analysis

In surface studies, one is confronted with the difficulty of detecting a small number of surface atoms in the presence of a large number of bulk atoms: a typical solid surface has 10^{15} atoms/cm^2 as compared with 10^{23} atoms/cm^3 in the bulk. In order to be able to probe the properties of solid surfaces using conventional methods, one needs the use of powders with very high surface-to-volume ratio so that surface effects become dominant. However, this technique suffers from the distinct disadvantage of an entirely uncontrolled surface structure and composition which are known to play an important role in surface chemical reactions. It is thus desirable to use specimens with well-defined surfaces which generally means small surface area, of the order of 1 cm^2, and examine them with tools that are surface sensitive.

It turns out that electrons with energies in the range of 10 to 500 eV are ideally suited for this purpose. Figure 4.1 shows a plot of the mean distance of electron penetration in solids as a function of the electron energy. The curve exhibits a broad minimum in the energy range between 10 and 500 eV, with the corresponding mean free path on the order of 4 to 20 Å. Electron emission from solids with energy in this range must therefore originate from the top few atomic layers. By extension, all experimental techniques involving the incidence onto and/or emergence from surfaces by electrons having energy between 10 and 500 eV are thus surface sensitive.

Many such techniques have been developed and used. Low-Energy Electron Diffraction, in which electrons are elastically scattered off a surface, has been the most successful among those for surface crystallography. Inelastically scattered electrons also

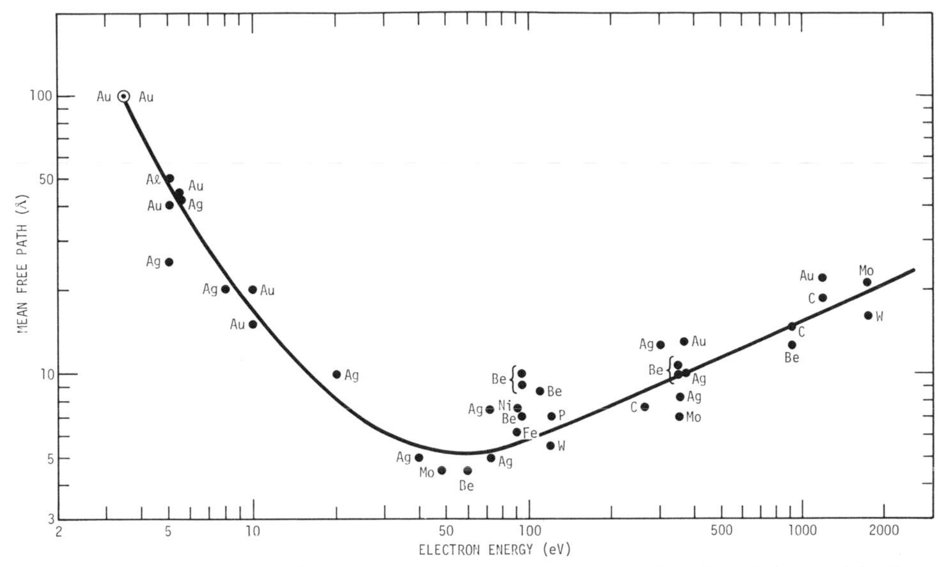

Fig. 4.1. The "universal curve" for the electron mean free path as a function of electron kinetic energy. Dots show individual measurements

provide surface-structural information in a method called High-Resolution Electron Energy Loss Spectroscopy. Secondary electrons ejected from the surface by incident electrons can also be used, especially for chemical composition analysis (in Auger Electron Spectroscopy), an important and essential function in surface investigations. Impinging light, ions or atoms can cause electrons to leave the surface in ways characteristic of the surface structure: thus photoemission, Ion Neutralization Spectroscopy, Surface Penning Ionization and many other techniques have been applied in surface analysis. Low-energy atomic and ionic scattering off surfaces are uniquely surface-sensitive processes in which the scattering particles only come into contact with the outermost surface atoms, since no penetration through atoms occurs. Light reflection can also be turned into a surface-sensitive tool, as in Infrared Spectroscopy.

It must be emphasized that all the surface analysis techniques developed so far have particular, often stringent, limitations. Thus some are more sensitive to chemical composition (AES), others to relative atomic positions (LEED, angle-resolved photoemission and SPI), others still to vibration modes (HREELS and IR), some to electronic levels (angle-integrated photoemission and INS), etc. To extricate the nature of any given surface it has become necessary to use several complementary and/or supporting techniques in parallel. For example, AES is a basic method used in nearly all studies, while in addition LEED together with Thermal Desorption Spectroscopy and photoemission or LEED together with HREELS and work function measurements might be used.

We now briefly describe the mechanism, capabilities and limitations of the main techniques used in surface analysis[1)], classifying them by the nature of the information obtained with them.

4 Methods Sensitive to Atomic Geometry at Surfaces

4.1 LEED (Low-Energy Electron Diffraction)

LEED has yielded a relatively large amount of structural information and will therefore be treated relatively extensively here[2].

In LEED, electrons of well-defined (but variable) energy and direction of propagation diffract off a crystal surface. Usually only the elastically diffracted electrons are considered and we shall do so here as well. The electrons are scattered mainly by the individual atom cores of the surface and produce, because of the quantum-mechanical wave nature of electrons, wave interferences that depend strongly on the relative atomic positions of the surface under examination.

The de Broglie wavelength of electrons, λ, is given by the formula λ (in Å) $= \sqrt{150/E}$, where E is measured in eV. In the energy range of 10 to 500 eV the wavelength then varies from 3.9 Å to 0.64 Å, smaller or equal to the interatomic distances in most circumstances. Thus the elastically scattered electrons can diffract to provide information about the periodic surface structure. The LEED experiment is carried out as follows: a monoenergetic beam of electrons (energy resolution approximately 0.2 eV) in the range of 10 to 500 eV is incident on one face of a single crystal. Roughly 1 to 5% of the incoming electrons are elastically scattered and this fraction is allowed to impinge on a fluorescent screen. If the crystal surface is well-ordered, the diffraction pattern consisting of bright, well-defined spots will be displayed on the screen. The sharpness and overall intensity of the spots is related to the degree of order on the surface. When the surface is less ordered the diffraction beams broaden and become less intense, while some diffuse brightness appears between the beams. A typical set of diffraction patterns from a well-ordered surface is shown in Fig. 4.2.

The electron beam source commonly used has a coherence width of about 100 Å. This means that sharp diffraction features are obtained only if the regions of well-ordered atoms ("domains") are of $(100\ Å)^2$ or larger. Diffraction from smaller size domains gives rise to beam broadening and finally to the disappearance of detectable diffraction from a disordered (liquid-like) surface.

One may distinguish between "two-dimensional" LEED and "three-dimensional" LEED. In two-dimensional LEED one observes only the shape of the diffraction pattern (as seen and easily photographed on a fluorescent screen). The bright spots appearing in this pattern correspond to the points of the two-dimensional reciprocal lattice belonging to the repetitive crystalline surface structure, i.e., they are a (reciprocal) map of the surface periodicities. They therefore inform us about size and orientation of the surface unit cell: this is important information, since the presence of, for example, reconstruction-induced and overlayer-induced superlattices is made immediately visible. This information also includes the presence or absence of regular steps in the surface[3]. The background in the diffraction pattern contains information about the nature of any disorder present on the surface[4]. As in the analogous case of X-ray crystallography, the two-dimensional LEED pattern in itself does not allow one to predict the internal geometry of the unit cell (although good guesses can sometimes be obtained): that

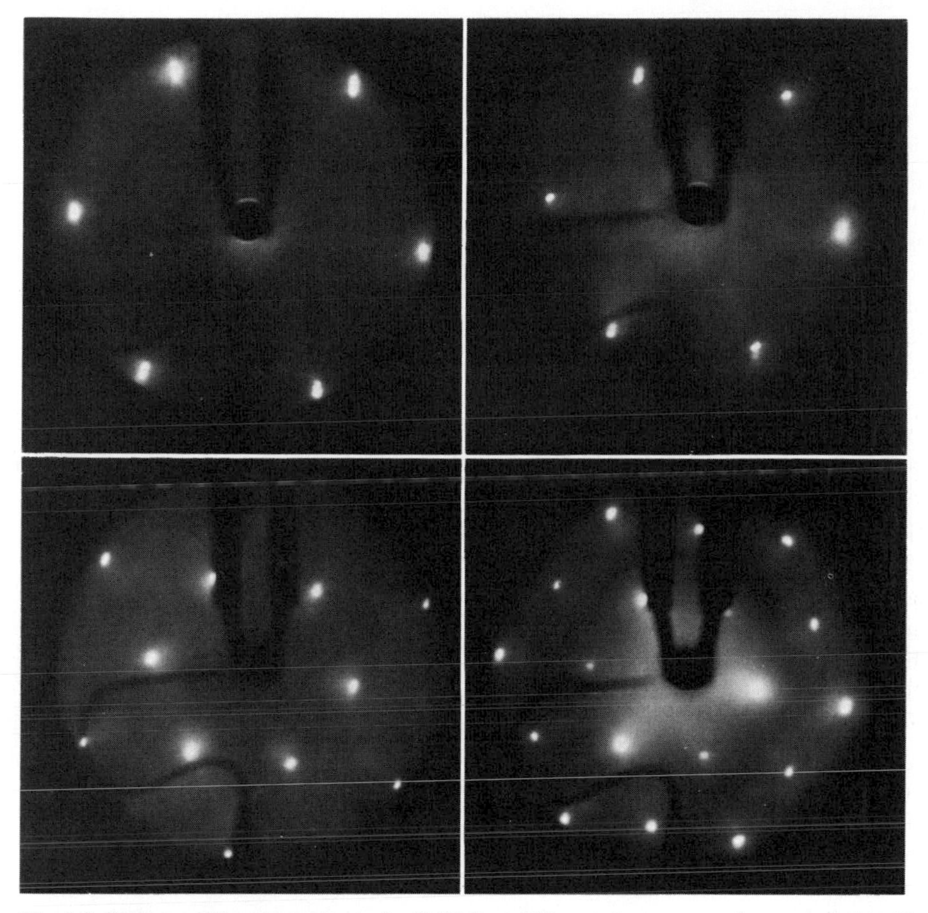

Fig. 4.2. Electron diffraction patterns for Pt(111) at different electron energies, at normal incidence. With increasing energy the diffraction spots converge toward the specular reflection spot, here hidden by the crystal sample

requires an analysis of the *intensities* of diffraction. Nevertheless, two-dimensional LEED already can give a very good idea of essential features of the surface geometry, in addition to those mentioned before. Thus one may follow the variation of the diffraction pattern as a function of exposure to foreign atoms: it is often possible to obtain semi-quantitative values for the coverage, for the attractive and/or repulsive interactions between adsorbates[5], for some details of island formation[6], etc. The variation of the diffraction pattern with changing surface temperature also provides information about these interactions [in particular at order/disorder transitions[6,7],] while the variation with electron energy is sensitive to quantities such as surface roughness perpendicular to the surface and step heights[3a].

In three-dimensional LEED, the two-dimensional pattern is supplemented by the intensities of the diffraction spots (thereby focusing the attention on the periodic part of the surface structure, i.e., the ordered regions) to investigate the three-dimensional internal structure of the unit cell. This is most readily done by considering the variation of the spot intensities as a function of electron energy and/or direction of incidence.

Measurements of the diffracted electron beam intensities can be carried out by various techniques that include photographing the fluorescent screen or collecting the electrons at any given angle of emission. The resultant intensity vs. electron energy curves (usually called I-V curves) or I-θ or I-ϕ curves (for variation of the polar or azimuthal incidence angles, respectively), serve as the basis for surface structural analysis. A set of I-V curves from a Pt(111) crystal face is displayed in Fig. 4.3. They exhibit pronounced peaks and valleys which are indicative of constructive and destructive interference of the electron beam scattered from atomic planes parallel to the surface as the electron wavelength is varied. Often, Bragg peaks (due to simple interference between electrons backscattered from different atomic planes, as in X-ray diffraction) can be identified. However, in addition to these and also overlapping with these, there are usually extra peaks that are due to multiple scattering of electrons through the surface lattice.

The presence of well-defined peaks and valleys in I-V curves indicates that LEED is indeed not a purely two-dimensional surface diffraction technique. There is a finite penetration and diffraction takes place in the first 3 to 5 atomic layers. The depth of penetration affects peak widths markedly: the shallower the penetration, the broader is the diffraction peak. By simulating such I-V curves numerically with the help of a suitable theory, it is often possible to determine the relative positions of surface atoms (including therefore bond lengths and bond angles)[2a,f]; it may also be possible to indicate roughly the thermal vibration state of surface atoms[2a]. However, a chemical identification of the surface atoms is not possible with LEED.

The analysis of LEED intensities requires a theory of the diffraction process. This is a nontrivial point because of the fact that multiple scattering of the LEED electrons by the surface is always present and is not easy to represent in a theory. Even a simple single-scattering theory (such as is used in X-ray crystallography) must justify its validity with respect of the neglect of multiple scattering. Clearly the computational effort increases with the complexity of the theory and so the present situation has arisen in which a variety of theories of different complexity co-exist, each justifying its existence by a different compromise between computational effort, ease of use, amount of experimental data required, reliability, accuracy, type of information produced and range of applicability.

In order of increasing computational complexity, the following main theoretical methods are used today in LEED:

- simple kinematical theory (simple s-wave, i.e., isotropic, scattering, as in X-ray diffraction theory, with inner potential correction);
- the above with an anisotropic atomic form factor;
- any of the above with averaging of experimental data (to average away multiple scattering effects)[8];
- any of the above with Fourier transformation from momentum space to coordinate space[9];

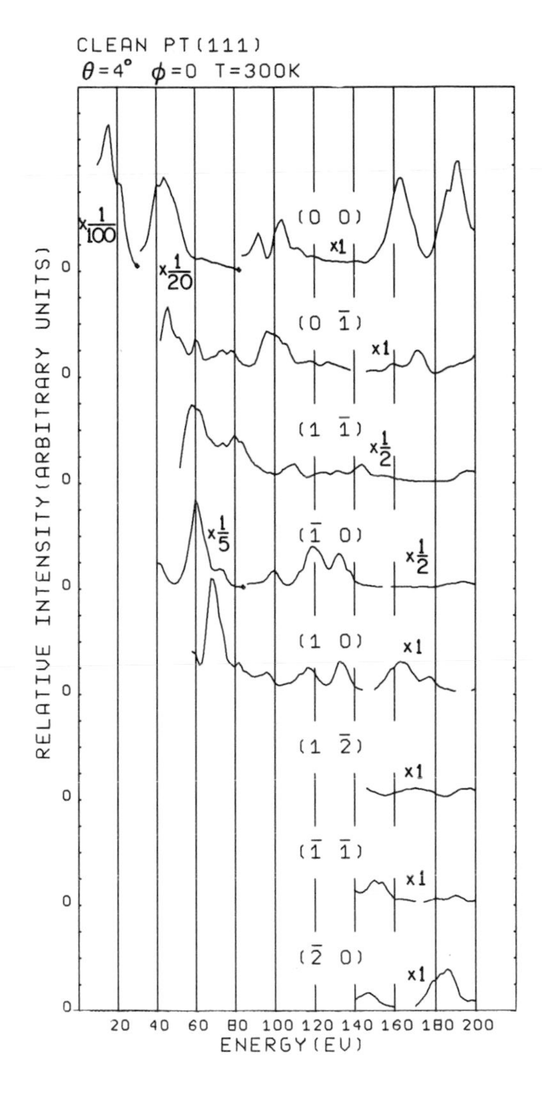

Fig. 4.3. Experimental intensity vs. voltage (energy) curves for electron diffraction from at Pt(111) surface. Beams are identified by different labels (h,k) representing reciprocal lattice vectors parallel to the surface. An incidence angle of 4° from the surface normal is used

- quasi-dynamical theory (this includes multiple scattering between, but not within atomic layers parallel to the surface)[10];
- iterative dynamical theory (multiple scattering is iterated to convergence; examples are the Renormalized Forward Scattering[2a] and Reverse Scattering Perturbation[11] methods);
- full dynamical theory[2a] (multiple scattering is included in closed form; examples are Beeby's matrix inversion, the Layer Doubling and the Bloch-wave methods; the first two of these assume a crystal of finite thickness, which is increased until convergence of the results in the case of Layer Doubling);
- Spin-polarized LEED theory[12] (relativistic spin-dependent effects are added to a dynamical theory);

– LEED theory for disordered surfaces[13,4c] (effects of disorder in the surface structure are added to a full dynamical theory).

Roughly speaking the LEED theories in the order listed above give increasing accuracy and reliability of the structural results. Few of these methods can produce a result with a single calculation: with most methods a trial-and-error search for the actual structure must be undertaken (giving rise to the desirability of independent hints about the structure from other surface analysis techniques). Most results of surface crystallography by LEED to date have been obtained with iterative or full dynamical theories. With these theories as they stand today the limitations on the possibilities are roughly the following:

– unit cell areas are limited to about 25 Å^2 (the equivalent of a (2×2) superstructure on a low-index face of a simple metal);
– the number of atoms in one unit cell per layer parallel to the surface is limited to about 4;
– the accuracy in distances perpendicular to the surface is, depending on the case, about 0.1 Å or better (for comparison, atomic vibration amplitudes at room temperature are usually of the order of 0.1 Å);
– the accuracy in distances parallel to the surface is of the order of 0.2 Å, unless a well-defined symmetrical atomic position can be assumed, in which case one assumes no uncertainty;
– the resulting accuracy in bond lengths varies from less than 0.05 Å (for bonds more or less parallel to the surface, assuming no uncertainty in the bonding site) to 0.2 Å; this translates to a relative uncertainty between less than 2 and 10% of the bond length.

In order to indicate the theoretical ideas involved in LEED crystallography, we now outline the main dynamical (i.e., multiple-scattering) methods used to compute I-V curves for comparison with the experiment (cf., also References 2a and 2f).

The crystal surface is imagined to consist of individual atomic layers parallel to the surface. Whenever convenient, the LEED electrons between these layers are represented by a set of plane waves (to each diffracted beam corresponds one plane wave), as the electron-solid interaction potential is assumed to be a constant between the layers. These plane waves are diffracted any number of times by these individual atomic layers, whose diffraction properties are discussed below and assumed known here. The multiple scattering between layers is treated usually in one of three ways:

1. In the Bloch-wave method the periodicity of the crystal perpendicular to the surface underneath the deviating surface region is exploited: the Bloch theorem applies and enables the electronic eigenfunctions (the Bloch waves) to be determined. These eigenfunctions are then matched across the surface region to the conditions outside of the surface (consisting of one incident beam and a set of reflected beams): this matching fixes the intensities of the reflected beams.

2. In the Layer Doubling method the diffraction properties of pairs of layers are determined exactly from those of the individual layers: this is done by summing up the multiple scattering between the layers as in a geometrical series, but using matrix inversion rather than the series expansion. By repeating this combination of layers, the

crystal can be built up layer by layer until convergence of the surface reflectivities (in the periodic bulk each step can double the thickness of the growing slab of layers). This procedure converges because of the presence of electron "absorption": most of the incoming electrons lose energy as they move through the crystal surface and are therefore removed from the flux of elastically scattered electrons that we are interested in (this absorption is simulated by a mean free path or by an imaginary part of the electron energy). The Layer Doubling method is computationally more efficient than the Bloch-wave method and it is more flexible in terms of varying the surface structure, as is necessary in a structural search.

3. In the Renormalized Forward Scattering (RFS) method substantial computation time is saved by recognizing and exploiting the fact that many multiple-scattering processes are too weak to contribute significantly to the diffracted intensities. Namely, *back*scattering off any atomic layer is usually weak (forward scattering is not) and therefore scattering paths are ordered according to increasing number of such backscatterings. This method is cast in a convenient iterative form, providing a most efficient computation scheme for interlayer multiple scattering (non-convergence, however, occurs in cases of very strong scattering and small interlayer spacings).

The individual layer diffraction properties needed as input to the methods described above are obtained as follows. The multiple scattering between the atoms of a given layer can be summed up exactly to produce a matrix inversion, in a way analogous to the treatment of interlayer multiple scattering in the Layer Doubling method. Computationally this is manageable only when the individual atoms of the crystal surface are assumed to be spherical and spherical waves may be used between the atoms. One therefore uses an electron-solid interaction potential consisting of non-overlapping spherically symmetrical regions with a constant interstitial value: such a potential is called a "muffin-tin potential." This approach is used in all current dynamical LEED computations for treating individual atomic layers. These layers may have more than one atom per unit cell and these atoms need not be coplanar.

In fact it is possible to consider the entire surface as a single thick layer composed of perhaps 5 individual layers (since only finite electron penetration occurs) that would be treated by this "matrix inversion method in angular momentum space": however this solution gives rise to matrix dimensions that rapidly exceed the possibilities of all existing computers; also the matrix inversion would have to be repeated for each different surface geometry, a waste that is largely overcome by using plane waves between atomic layers. Generally speaking, plane waves are used as often as possible, because they offer clear computational advantages.

A perturbation expansion version of this matrix inversion method in angular momentum space has been introduced with the Reverse Scattering Perturbation (RSP) method, in which the ideas of the RFS method are used: the matrix inversion is replaced by an iterative, convergent expansion that exploits the weakness of the electron backscattering by any atom and sums over significant multiple scattering paths only. This method can be applied to individual atomic layers or to a thick layer representing the entire surface, but remains relatively time-consuming compared to the plane-wave methods, except when the separation between individual layers becomes small ($\lesssim 0.5$ Å).

To obtain the above-mentioned *layer* diffraction properties, one needs as input the single-atom scattering properties. These, in the case of spherically symmetrical atoms, are given by a set of phase shifts (which are species- and energy-dependent). The phase shifts are obtained by a numerical integration of the Schrödinger equation with an atomic potential, that in turn has to be generated from first principles, including electrostatic and exchange-correlation effects in the electron-atom interaction, as well as the effects of neighboring atoms.

In practice the atomic scattering properties are modified by including a Debye-Waller factor that represents the effect of the thermal vibrations of the surface atoms: thus the temperature correction is applied at each scattering in each chain of scatterings. Thermal atomic vibrations have in LEED an effect similar to that in X-ray diffraction: the intensity of the diffracted beams is decreased, while the background reflection between beams increases (electron-phonon scattering can impart a change of momentum parallel to the surface that generates intensity in directions other than those of the beams). The decrease in intensity of the beams also behaves often as in X-ray diffraction in spite of multiple scattering: an exponential decrease is usually observed, which can be described by the Debye-Waller factor

$$\exp\left(\frac{-3\,|\Delta\vec{k}\,|^2 T}{m k_B \theta_D{}^2}\right)$$

where Δk is the change in electron momentum, T the temperature, m the atomic mass and k_B Boltzmann's constant. This factor involves a material-dependent constant, the Debye temperature θ_D, that quantifies the rigidity of the crystal lattice and thereby influences the amplitudes of the vibrations. θ_D can be obtained experimentally for a surface through measurement of the Debye-Waller factor and typical results are shown in Fig. 4.4: the experimental θ_D is observed to vary irregularly with the LEED electron energy, tending at high energies to a constant that lies in the neighborhood of the value of θ_D for the inside of the crystal. The Debye temperature θ_D, being a material constant, should not depend on characteristics of the probe used to measure it (here the electron energy). The interpretation of this anomalous behavior is the following. At high energies the electrons penetrate deeply into the surface and sample the bulk properties of the crystal. As the energy is lowered, two effects are noticed. First, multiple scattering produces rapid variations in the experimental θ_D with electron energy. Some decrease in θ_D is also attributable to multiple scattering at low energies, but not enough. This second effect (a decrease in θ_D at lower electron energies) is to a large extent explained by larger thermal vibration amplitudes of surface atoms as compared with bulk atoms, combined with the shallower penetration of low-energy electrons. One can estimate in this way that clean-surface atoms have vibrations enhanced by typically 50% in the direction perpendicular to the surface, and little enhanced in directions parallel to the surface, in good agreement with theoretical predictions. Due to the complication of multiple scattering, however, attempts to extract more precise information about surface vibrations have not met with success, not even in the simplest case, that of clean metal surfaces.

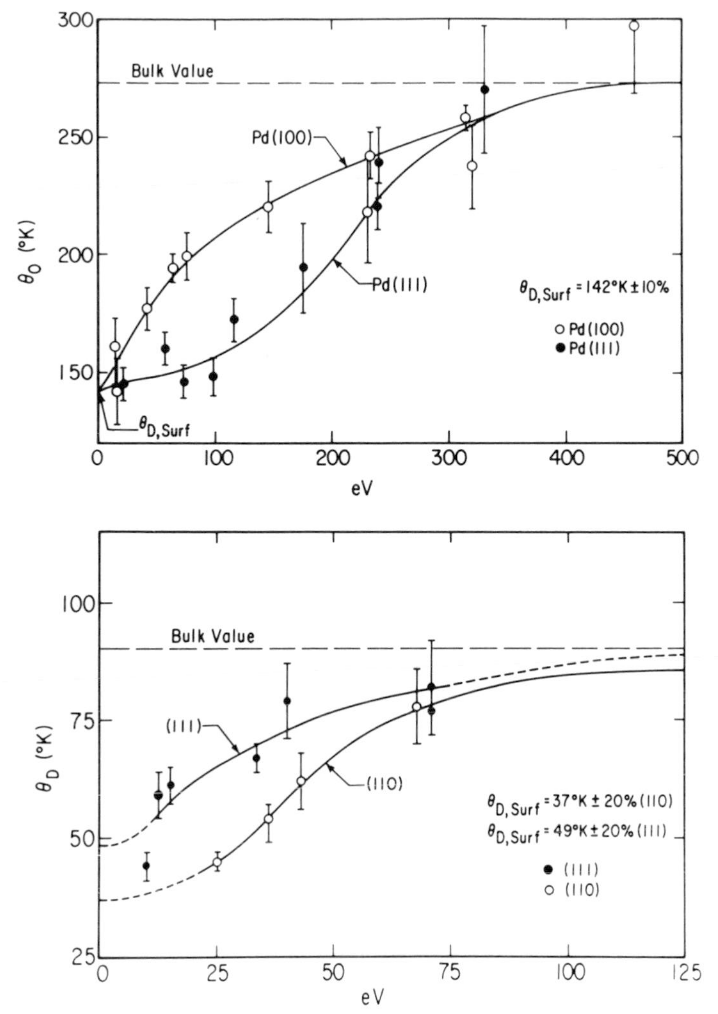

Fig. 4.4. Measured Debye temperatures as a function of electron kinetic energy for Pd (*top*) and Pb (*bottom*) surfaces

4.2 RHEED and MEED (Reflection High-Energy Electron Diffraction and Medium-Energy Electron Diffraction)

RHEED and MEED[14)] differ from LEED in the range of energies used: while in LEED "low" energies of about 10-500 eV are used, in RHEED "high" energies of about 1–10 keV are used, with MEED bridging the intermediate energy range. The surface sensitivity of LEED is guaranteed by the small mean free path (~5–10 Å) at the low energies. At higher energies the mean free path increases (~20–100 Å for RHEED energies) and so impairs the surface sensitivity on the atomic scale, unless grazing angles of incidence and emergence are used: this is therefore the normal choice in MEED and

RHEED. However, grazing angles of incidence put stronger requirements on the large-scale planarity of the surface than the roughly perpendicular incidence directions used in LEED. The multiple scattering present in LEED is also present at the higher energies: corresponding dynamical theories have been developed,[14b,c] but there is a lack of accurate experimental data to interpret.

In many studies the chemisorption and the surface reaction is just the first step in a series of solid state reactions that take place as atoms from the surface move into the bulk. Corrosion, oxide, carbide and other compound formations are generally initiated at the surface and then propagate into the bulk. There may be a concentration gradient of certain constituents at the surface in a multicomponent system that would influence the mechanical or chemical properties of the system. Hardening of materials and other forms of passivation treatment frequently involve introduction of certain substances only in the near surface region. For the investigation of these problems RHEED is a powerful technique.

4.3 Electron Microscopy

The electron-optical techniques are increasingly often being applied under conditions of ultra high vacuum, allowing the study of surfaces under controlled circumstances[15]. The most significant developments for electron microscopy however have been in the imaging mode where considerable enhancement of resolution has occurred, leading to various forms of operation.

Transmission electron microscopy (TEM) is most like light-optical microscopy. Electrons, accelerated in the 100 kV range from either a thermionic or field emission gun, are used to illuminate a specimen which is typically 3 mm in diameter and $\leqslant$ 5000 Å thick. After transmission through the specimen, the electrons are focused by an electromagnetic objective lens to form an image. Other lenses are also utilized in the optical column to demagnify the source onto the specimen, to image the diffraction pattern at the back focal plane of the objective and to magnify up to 1 million times the image produced by the objective. The final image is recorded photographically and can show clearly resolved detail in the 2 to 3 Å range.

Even though the TEM image is a two-dimensional projection of the specimen structure, it is highly sensitive to changes in thickness. Therefore, particularly when enhanced by certain imaging techniques, surface steps and morphological irregularities can be examined visually. These imaging techniques employ either diffraction contrast for thick specimens ($\gtrsim$ 1500 Å) or defocus contrast for thinner specimens ($\geqslant$ 500 Å). Step structures having a height of $\sim$ 10 Å may be revealed by either method, although the latter technique has successfully been used in the imaging of monolayer steps and of single, heavy-element atoms on specially prepared substrates.

Scanning electron microscopy (SEM) utilizes a highly focused electron beam which is scanned over the surface of the specimen. Since penetration through the specimen is not essential for this instrument, thicker samples (cm range) and lower accelerating potentials (low kV range) are commonly used. The most popular mode of operation is the emissive mode which utilizes those electrons that have either been emitted by the specimen as secondaries, or have been backscattered. Due to the strong dependence of the number of collected electrons on incident illumination angle, surface topography is dramatically revealed by this technique. Resolution is primarily determined by the spot size of the focused electron beam, and is rather less than 100 Å. With thin enough specimens, a detector may be placed such that it collects the transmitted electron signal. This is the principle behind scanning transmission electron microscopy (STEM), which has allowed resolution down to 2–3 Å as a result of very small electron probe sizes. The electron-atom cross sections here also limit observability to heavy atoms on light substrates (typically carbon films). The thin substrate films moreover do not have single-crystal surfaces, but are amorphous.

In electron microscopy, generally intense electron beams are used which can severely damage the surface (however, at high energies the electron-atom cross sections become smaller). Often, large magnetic fields are also present that could affect the surface structure.

Since the first electron-microscopical observation of a heavy atom on a surface[15a], different studies have looked at effects related to individual atomic adsorbates. These include diffusion along the surface (atoms can be tracked in real time), giving results in agreement with equivalent FIM observations, and pair spacing distributions, showing for example a peak in the distribution near 4-5 Å for uranium atoms on a carbon surface[15c]. Clustering can be studied in some cases as well.

It would be interesting to extend such studies to other light-atom substrates, such as the metals beryllium and aluminium, and to investigate step effects. Heavier-atom surfaces can also be analyzed in the form of thin films of mono-atomic thickness on a lighter substrate, as has recently been done[15d].

4.4 FIM (Field Ion Microscopy)

In Field Ion Microscopy[16], a hemispherical sample tip is imaged by allowing a gas (usually helium, but also hydrogen, neon and others) to ionize at the surface of the tip under the influence of a strong applied electric field, which also projects the ions onto a screen that can be photographed. The ionization probability depends strongly on the local field variations induced by the atomic structure of the surface: protruding atoms generate appreciably stronger ionization than atoms embedded in close-packed atomic planes and so produce individual bright spots on the screen. The imaging by the ions from the sample tip to the screen occurs with very little motion tangential to the tip surface, especially at low temperatures ($T \sim 21$ K is often used for that reason), allowing a resolution of 2–3 Å. The use of small radius tips (500–2000 Å) is needed to produce the large field required for ionization, but also is responsible for the immense magnification of this microscope: the tip surface is directly imaged with a magnification of about 10^7.

Only a limited class of materials withstands the strong electric field at the sample tip without desorption of the surface occurring. Thus, mainly metals with large atomic number (W, Pt, Rh, Re, for example) are used. Any adsorbates are equally affected by field desorption, greatly restricting the range of usable substrate-adsorbate systems that can be studied by FIM. Furthermore, the properties of surfaces under high-electric-field conditions may differ from those of the field-free state. Nevertheless, FIM has been very helpful in understanding the properties of metal surfaces and metal-on-metal surfaces, including defect structures, thermal disordering, atom-atom interactions, two-dimensional cluster formation and evolution, and atomic surface diffusion (since real-time observation of individual atoms is possible). As an example, we may cite the study by FIM of the self-diffusion and correlated motions of Rh atoms on Rh substrates[16f)]. The self-diffusion shows strong anisotropies on crystal faces that are channelled, such as fcc(110), with a large mobility along the channels and a low mobility across the channels.

Concerning the analysis of the detailed geometrical surface structure, FIM unfortunately does not provide the depth information required to investigate the coordination of surface atoms to underlying atoms (i.e., layer registries cannot be determined) or to measure bond lengths.

4.5 LEIS, MEIS and HEIS (Low-, Medium- and High-Energy Ion Scattering)

Ion scattering at low energies ($\lesssim 2$ keV), medium energies ($\sim$ 50–500 keV) and high energies ($\gtrsim 500$ keV) has been used to study surface structures[17)]. In LEIS[17a)] high surface specificity is obtained because the very large cross sections for ion scattering ensures scattering off the outermost atomic layer only. Mutual shadowing of surface atoms is exploited to investigate their mutual positions. However, the physics of the scattering process at these energies is not well understood, leading to uncertainties in position determinations of the order of 0.5 Å. This does allow gross (but important) observations such as whether adsorbed atoms lie tucked away between substrate atoms [for example, in the channels of fcc(110) surfaces] or instead are situated in more exposed positions [for example, on the ridges of fcc(110) surfaces]. At high energies[17c)], attained by the use of ion accelerators, the cross sections become very small, allowing deep penetration into the surface. But surface sensitivity is maintained by using channeling (penetration along open channels in the bulk crystalline structure) and looking for the blocking of this channeling by surface atoms whose positions deviate from the bulk positions. This amounts to a kind of "triangulation", in which the directions of the lines connecting pairs of surface atoms are identified by looking for shadowing of one atom of each pair by the other atom: these directions are then sufficient information to determine the relative positions of surface atoms. At high energies the well-understood Rutherford scattering is the dominant mechanism, simplifying the interpretation. At medium energies ion scattering gives additional information about surface composition, since backscattered ions lose an amount of energy that depends on the mass of the hit surface atom[17b)]. A general problem with ion scattering is that the thermal vibrations of the surface atoms complicate the interpretation. Sometimes computer simulations of the scattering are therefore made to sort out the structural from the thermal effects. In principle accuracies of better than 0.1 Å in the determination of atomic positions are possible.

As an example of such work, the clean Ni(110) surface has been studied with MEIS[17d)], confirming the contraction of the topmost interlayer spacing observed by LEED. Adsorption of half a monolayer of sulfur was found to cancel that contraction and actually expand the topmost interlayer spacing beyond its bulk value. The position of the adsorbed sulfur in the deepest hollows of the substrate surface could also be determined, the result being in agreement with that of a previous LEED study (although there a spacing expansion was not investigated).

4.6 Atomic Scattering and Diffraction

The de Broglie wavelength associated with helium atoms is given by

$$\lambda(\mathrm{A}) = \frac{h}{(2ME)^{1/2}} = \frac{0.14}{E(eV)^{1/2}}$$

Thus atoms with thermal energy of about 0.02 eV have $\lambda = 1$ Å and can readily diffract from surfaces. A beam of atoms is chopped with a variable frequency chopper before striking the surface. This way, an alternating intensity beam signal is generated at the mass spectrometer detector, that is readily separated from the "noise" due to helium atoms in the background.

There are three processes observed during the scattering of atomic beams[18)] of helium that are displayed in Fig. 4.5. There is specular reflection of the helium atoms from the surface (i.e., the helium atoms scatter at an angle that is equal to the angle of incidence) (*curve a*). There is rainbow scattering (*curve b*) that results in the appearance of multiple peaks; this type of scattering may be viewed as the classical limit of diffraction: it is due to scattering of atoms by the varying surface potential. The third type of process (*curve c*) is diffraction, and it appears to be detectable from suitable surfaces for values of $d > \lambda > 0.15\ d$, where d is the interatomic distance in the surface, and when using fairly monochromatic incident atomic beams.

All three of these processes have been observed under various conditions of the beam scattering experiment. A typical diffraction pattern of helium from a LiF(100) surface is shown in Fig. 4.6. The first order beams, although broad, are clearly discernible. Diffraction from LiF has also been observed using neon, hydrogen and deuterium. Diffraction of helium from other surfaces – tungsten(112), silver(111) and tungsten carbide and silicon(111) – has also been detected. Another phenomenon, the presence of bound states for the incident helium atom has also been detected during the helium-lithium-fluoride diffraction experiments. A fraction of the helium atoms appear to be trapped at the surface in weakly bound (2 to 12 cal) states but can readily translate along the surface before re-emission without energy loss. Such a phenomenon has been predicted by Lennard-Jones and Devonshire in 1938.

Rainbow scattering has been detected from high Miller Index stepped platinum surfaces. Typical rainbow scattering patterns are shown in Fig. 4.7. The increase in intensity of the surface rainbows, as displayed by this figure, for an increase in the angle of incidence, qualitatively follows the trend predicted from calculations by McClure

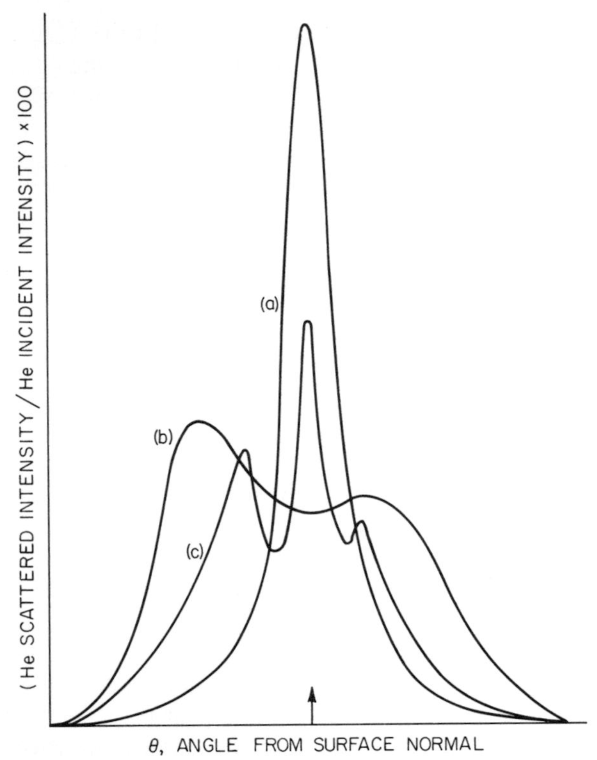

Fig. 4.5. Schematic atomic scattering distributions for: (*a*) specular scattering, (*b*) rainbow scattering and (*c*) diffraction

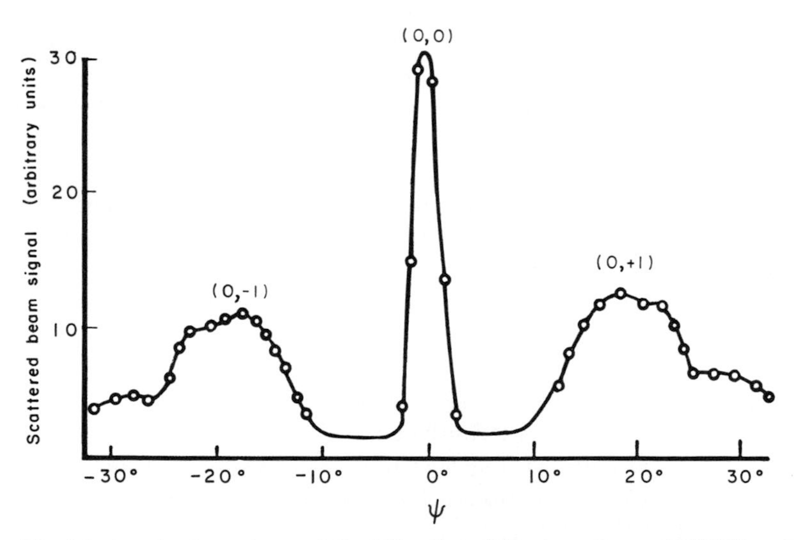

Fig. 4.6. Angular dependence of the diffraction of He atoms from a LiF(100) surface

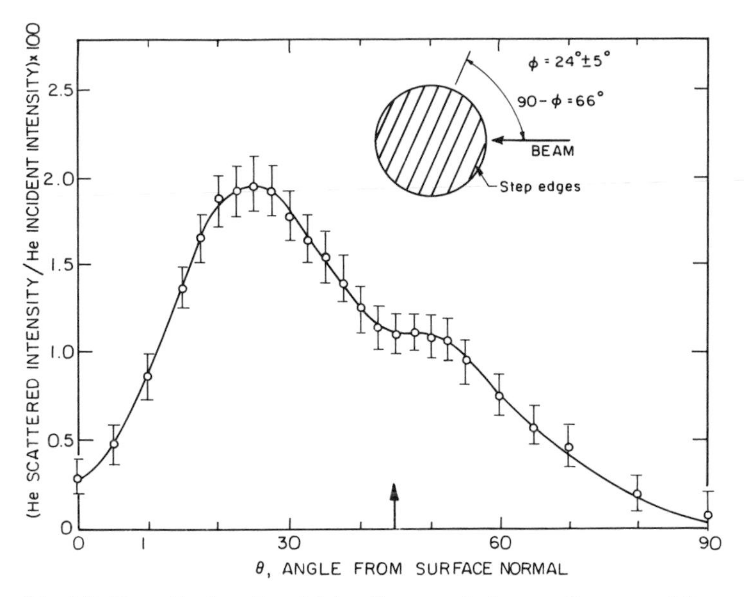

Fig. 4.7. Normalized scattered intensity vs. angle from surface normal for an azimuthal angle of $\phi = 24^\circ$ with a fixed angle of incidence of 45° on a stepped Pt(553) surface

for classical scattering for helium atoms. At grazing angles of incidence, the surface appears less spatially rough and thus allows a more intense elastic scattering contribution. At more normal angles of incidence, the elastic scattering distribution is less intense due to an increase in spatial roughening of the surface as seen by the helium atom. Rainbow scattering has been observed from a stepped platinum surface while the smooth Pt(111) surface exhibits only specular scattering.

Helium, in general, gives strong specular scattering from ordered surfaces. As the surface temperature is increased, the specular beam intensity drops. This effect is similar to that observed for low-energy electrons and for X-rays and is due to surface atom vibrations that give rise to the Debye-Waller factor. However, the form of the Debye-Waller factor is different for atom beam diffraction as compared to the scattering of these other two surface probes. The slow, low energy helium atoms encounter the attractive surface potential that has a well depth similar to the thermal energy of approaching atoms and the temperature dependence of the scattering is influenced by the well depth of this potential. Beeby has derived a formula for the temperature dependence of the helium beam intensities:

$$I = I_0 \exp\left(-\frac{25 M_g T D}{M_s \theta_s^2} \right)$$

Here D is the depth of the atomic potential as sensed by the incident atom and M_g and M_s are the masses of the scattering atoms and surface atoms, respectively, and θ_s is the surface Debye temperature. For the diffracting X-rays and electrons that have much higher energy as compared to helium atoms, the surface potential well can be neglected as having an unimportant effect on scattering. It appears, therefore, that the temperature dependence of atom scattering can yield information on the attractive potential that is operative during the solid-gas interaction.

Another important property of the specularly scattered fraction of atoms is their great sensitivity to surface disorder. On scattering from a well ordered surface, nearly 15% of the scattered helium atoms appear in the specular helium beam. This fraction decreases to 1 to 5% when the surface is disordered. Thus measurements of the fraction of specularly scattered helium can provide information on the degree of atomic disorder in the solid surface.

4.7 SEXAFS (Surface-Sensitive Extended X-Ray Absorption Fine Structure)

In SEXAFS incident X-rays of variable energy eject, for example, low-energy adsorbate core Auger electrons[19)], which by their small mean free path guarantee surface sensitivity on the atomic scale. It is possible to focus attention on adsorbates, for example, by considering only adsorbate Auger lines: it is sufficient that the adsorbate be different from substrate atoms and that suitable Auger lines are present. The yield of the ejected electrons is modulated as a function of incident energy due to interference between outgoing electrons and electrons backscattered from the neighboring atoms. By Fourier-analysis of this modulation, the interatomic distances can be extracted with an accuracy that may exceed that of LEED. It is also possible to find the number of neighbors at the individual distances, thereby fixing the adsorption site through the coordination number[19b)].

One of the basic physical inputs to the SEXAFS analysis is a set of phase shifts for electron scattering off surface atoms. The uncertainty in these is one of the limiting factors in the accuracy of the method. However, in some cases comparison with experimental EXAFS data from bulk material can help in circumventing the phase shift uncertainty: essentially the bulk and surface phase shifts are assumed equal and these then divide out in the ratio of the surface to bulk data.

A drawback of SEXAFS is that synchrotron radiation is needed as a source of X-rays.

5 Methods Sensitive to Electronic Structure and Geometry of Surfaces

5.1 UPS and XPS (Ultraviolet and X-Ray Photoemission Spectroscopy) and ARUPS (Angle-Resolved UPS)

Photon-induced emission of electrons is an obvious tool for structural analysis in two ways. Firstly, it is sensitive to the initial density of states of the emitted electrons (originating from the first few atomic layers of a surface), and so to the surface geometry. Secondly, if the angular distribution of the emitted electrons is considered, additional information about the initial electron states (in particular orbital shape and bonding symmetry) and about the atomic positions (through multiple scattering in the final electron state) can be obtained. UPS and XPS (XPS was at first called ESCA: Electron Spectroscopy for Chemical Analysis) differ in the photon energy range used. With UPS energies between a few eV and about 100 eV are chosen: this provides sensitivity to the valence electrons, the conduction electrons and the first deeper-lying bound states, all of which give access to information about the surface structure. In XPS higher photon energies (X-rays) are used, allowing deep core levels to be explored: these allow chemical identification (therefore the name ESCA), but they are of additional interest because they exhibit energy shifts ("chemical shifts") that are characteristic of the atomic environment.

UPS and XPS have been extensively used in surface analysis and have yielded much qualitative information about the surface geometry[20)]. Thus one can distinguish with these techniques between atoms that are in the adsorbed state, the same atoms that have penetrated the surface to form compounds and those that have remained in intermediate stages (often interpreted as incorporation within the topmost surface layer), since such differences appear as shifts in initial-state levels. It is also possible to investigate, with more or less precision, the adsorption orientation and the structural modifications of molecules deposited on surfaces. Thus CO adsorbed on metals has been extensively studied with UPS. It is easy to distinguish between molecularly adsorbed and dissociatively adsorbed CO: the characteristic molecular 4σ, 1π and 5σ levels are either present or absent, respectively, in the UPS spectrum (even if often the 5σ and 1π levels coincide in energy). In the case of molecular adsorption it is furthermore usually clear that the 5σ orbital, which is located more towards the C end of the molecule, undergoes a larger energy shift upon adsorption than the 4σ orbital, which is located towards the other end of the molecule. This strongly suggests that the molecule is bonded by its C end to the surface, with the O end sticking out away from the surface.

While little theory is needed in the above kind of analysis with UPS and XPS (unless a detailed understanding of "relaxation energies", and the like is sought[21)], the situation is quite different in ARUPS. It is now well-established that final-state multiple scattering effects are important there: a treatment of these processes along the lines of LEED theory (requiring well-ordered surfaces) is needed and is starting to produce encouraging results[22)]. Also needed are an adequate treatment of the initial state and the initial-to-final-state matrix elements and maybe of the refraction of the incident photons at the surface. It appears that ARUPS is sensitive to bonding symmetry more than to bond lengths[22e)] (LEED has the reverse trend) and that an analysis of initial states (such as bonding orbitals and surface states) is indeed possible. ARUPS has been applied successfully to clean metal surfaces, where for example the d-band emission could be reasonably well reproduced. It has also been applied to atomic overlayers [such as S and Se on Ni(100)], confirming the binding site determined previously by LEED, as well as to molecular adsorption of CO on Ni(100): the CO is confirmed to adsorb by its C end to a single nickel atom, as predicted by LEED, but some uncertainty as to the orientation of the molecular axis (normal to the surface or tilted by about 30° from that) remains.

The computational effort is larger in ARUPS than in LEED, since more physical processes are involved, so that for the limited purpose of surface crystallography LEED seems more appropriate. Furthermore, ARUPS is best done with synchrotron radiation, which limits its availability.

5.2 INS (Ion Neutralization Spectroscopy)

In INS[23)] slow, positively ionized noble gas atoms (typically He^+ ions) are allowed to neutralize at a surface by attracting surface electrons. The energy liberated is transferred to other surface electrons (not belonging to the incoming ion) which can leave the surface and be detected. The probability of this two-electron process involves the self-convolution of the surface density of occupied states, the density of final states and matrix elements for electron tunneling to the ion and for ejection of the detected electrons (a final-state problem also encountered in LEED, SPI and photoemission). The procedure is to extract by deconvolution from the measured emission probabilities the surface density of occupied states for energies between the Fermi level and about 10 eV below that, and to predict the relative atomic positions from that information about the electronic structure of the surface: for example, adsorbate-induced peaks will occur, that depend on the adsorbate and its position, as in UPS. This technique is primarily sensitive to the outermost atoms of the surface, in particular adsorbates, since the emitted electrons originate from those regions only. The difficulties in deconvoluting and interpreting the density-of-states information have limited the use of INS.

The technique has however been applied to the adsorption of O, S and Se on Ni(100), Ni(110) and Ni(111). In the case of O on Ni(100) a substrate reconstruction was inferred, with penetration of the adsorbate into the topmost substrate layer, in disagreement with LEED results. A small distortion of the substrate was concluded for S on Ni(100), but the same adsorption site was found as with LEED (cf. Sect. VI).

5.3 SPI (Surface Penning Ionization)

A metastable helium atom incident (with thermal kinetic energy) onto a surface allows a surface electron to tunnel to the unoccupied low helium level, enabling the excited helium electron to be emitted and detected[24)]. This occurs with energy and angular distributions characteristic of the electronic structure and therefore geometry of the surface, in a way similar to photoemission (see the heading: UPS and XPS). Whereas in photoemission the excitation occurs via a dipolar photon-electron interaction, in SPI it is an $|\vec{r}_1 - \vec{r}_2|^{-1}$ electron-electron interaction that takes place, so that the selection rules are different. A further difference appears in the surface sensitivity: the photoemission excitation is sensitive to the entire region covered by the initial state of the electron to be excited, while the SPI excitation occurs where the empty incident-atom wave function starts to overlap with the surface wavefunctions, i.e., at the outer edge of the outermost atoms (final-state multiple scattering of the emitted electron is however similar in SPI and photoemission). SPI has similarities with INS, but is intrinsically easier to interpret, mainly because of the absence of a deconvolution of the surface density of occupied states; furthermore SPI is more surface-sensitive even than INS.

SPI is currently being applied to clean metal surfaces and to CO adsorbed on such surfaces.

6 Method Sensitive to Electron Distribution at Surfaces

6.1 Work Function Measurements

The work function is the potential that an electron at the Fermi level must overcome to reach the level of zero kinetic energy in the vacuum. It is due to the interaction of an electron with the charges of the surface, through electrostatic, exchange and correlation effects[25,1b)]. Any change of charge distribution at the surface will in general change the work function: especially charge redistributions perpendicular to the surface can produce sizeable work function changes (up to 2–3 eV in some cases), as is for example common when an adsorbed layer is deposited on a surface. Therefore work function measurements have become a sensitive technique for monitoring the state of the surface, usually as a function of coverage.

Experimentally, the work function itself can be measured with, among other methods, photoemission, since the work function appears as a clearly distinguishable threshold energy there. Changes in work function are often measured by the Kelvin method, which uses a vibrating capacitor.

Unfortunately, the work function is a rather complicated (and not fully understood) function of the surface composition and geometry. The work function change is usually attributed to the formation of a dipole layer on the surface, such as occurs when charge flows from a substrate to an adsorbate, or vice versa. If σ is the dipolar charge density, d the dipole length (perpendicular to the surface) and e the electronic charge, then one can write

$$\Delta\phi = 4\pi e \sigma d.$$

This relation enables one to estimate the charge transfer from measured values of $\Delta\phi$ and values of d determined for example by LEED. (However, the picture of dipoles consisting of two point charges a distance d apart is a drastic simplification of the actual continuous charge distribution at a surface). This use of work function measurements is applied in Sect. VI to the case of atomic adsorbates.

The work function change is not normally proportional to the concentration of adsorbates, except at very low coverages ($\theta < 0.1$). The main reason is that dipoles mutually depolarize each other, the more so the higher the concentration. This effect is included in the following relation[25c)], based on considering the polarizability of the adsorbates,

$$\Delta\phi = \frac{4\pi e \sigma_0 \theta}{1 + 9\alpha\theta^{2/3}}$$

where σ_0 is the dipolar charge density at $\theta = 1$ and α is proportional to the polarizability. This relation only rarely fits the experiment, however, because the charge distribution at surfaces is governed by more complicated processes than polarizability, such as charge transfer between substrate and adsorbate. Thus, there are examples of sign reversal of the work function change as coverage varies, while there is no detectable change in adsorption geometry (cf. Sect. VI).

Nevertheless, general systematic observations are quite helpful. For example, the sign of the work function change for atomic adsorption is mostly that implied by the sign of the valency of the adsorbates, as one would expect; furthermore, in atomic adsorption it appears that large atoms (such as alkali atoms) produce large work function changes, mainly because of the size effect. Many organic molecules adsorbed on metal surfaces produce a decrease in work function, indicating the transfer of electrons from the molecules to the substrate; also π-bonding is often implied for such molecules.

The most common usage of work function changes is in the monitoring of the various stages of adsorption as a function of coverage: often the work function change will go through a minimum or maximum at particular coverages corresponding to the completion of, say, an ordered c(2×2) arrangement; also the onset of adsorption in new adsorption sites may be detected in this way.

7 Methods Sensitive to Chemical Composition at Surfaces

7.1 AES (Auger Electron Spectroscopy)

In AES surface atoms are ionized (usually by incident electrons) in their deep-lying electronic levels; these levels are then filled by electrons from higher-lying levels; the gain of energy either is emitted as X-rays or (in the AES process) it is transferred to electrons from other electronic levels, which are then emitted and detected at energies characteristic of the particular levels involved. Since the level energies depend strongly on the chemical identity of the atoms, the emitted electrons have energies characteristic of the chemical identity of the emitting species[26]. Each species has therefore a particular fingerprint in AES, which makes this technique ideal for the determination of the chemical surface composition. Thus AES is routinely used to monitor surface cleanliness in most surface analysis work: impurity concentrations of the order of 1% of a monolayer are detectable (except with hydrogen and helium, to which AES is not sensitive for lack of suitable energy levels). Similarly the surface composition of alloys can be investigated with AES. (By depth profiling, through the sputtering away of surface layers, this composition can be determined to any depth.) Such applications however do not clearly distinguish between atoms adsorbed on and atoms incorporated in the substrate, because the signal comes from a surface layer of thickness equal to the electronic mean free path, which is of the order of several atomic layers. The signal strength may be due to a small concentration of adsorbed atoms or to a larger concentration of atoms incorporated in the first few layers of the substrate. This ambiguity, coupled with uncertainties in the absolute yield of the AES excitation process for different species, makes a quantitative analysis of atomic concentrations at surfaces difficult and unreliable.

A different aspect of AES concerns shifts in the observed peak energies that are due to chemical shifts of atomic core levels. Thus one is able to distinguish between atoms in different chemical environments (in a way analogous to XPS). In particular, studies of different oxidation states of oxygen adsorbed on metals have shown chemical shifts that grow with increasing oxygen valency.

7.2 TDS (Thermal Desorption Spectroscopy)

The technique of TDS[27)] (also called flash desorption) measures the rate of desorption of adsorbed atoms and molecules from surfaces as the temperature is steadily raised; this is usually repeated at a succession of different coverages of the adsorbate. One or more broad peaks appear in the desorption rate vs. temperature curves. When more than one peak is observed, each peak corresponds to a different bonding state, since different temperatures are required to induce desorption from these states. One therefore can monitor with TDS how many different bonding states are occupied at any given coverage and one may also estimate their relative populations. Thus CO was discovered to have two different adsorption states, labelled α and β, on a W substrate. For both CO and H_2 on Pt(111) one state is found, while the same adsorbates on a stepped Pt(111) substrate exhibit two states: the new state is presumably to be attributed to adsorption at the steps. One often finds that with rising coverage first one state (site) is populated and then a second state (site) begins filling up.

The positions and relative heights of the desorption peaks as a function of initial coverage can be analyzed with appropriate models: one obtains in this way good estimates for the binding energies (heats of adsorption).

Sometimes also the coverage itself can be obtained from the TDS spectra, since a suitable integration over the spectra should give the amount of adsorbate removed.

The reliability of TDS is somewhat limited by the fact that the effect on the desorption of mutual interaction energies between adsorbates is probably substantial and poorly understood.

7.3 SIMS (Secondary Ion Mass Spectroscopy)

SIMS analyzes the mass and angular distribution of clusters of atoms ejected during ion bombardment of a surface[28)], using incident ions with kinetic energies of the order of 1 keV; the ejected clusters are often ionized. The knowledge of the cluster masses allows the chemical composition of the surface to be investigated. For example, a Ni(100) surface saturated with oxygen and bombarded with 2000 eV Ar^+ ions produces mainly clusters of Ni^+, Ni_2^+, Ni_3^+, $O^\pm$, $O_2^\pm$, $NiO^\pm$, Ni_2O^+, NiO_2^- and $Ni_2O_3^-$.

This list incidentally illustrates that the depth resolution of SIMS is of the order of a few atomic layers.

A quantitative surface compositional analysis requires the comparison of the experimental yield of the individual clusters with corresponding yields obtained theoretically: this may be done by numerical simulation of the complex collision process[28b)], but the accuracy of the result cannot yet be ascertained. The accuracy of the compositional analysis depends to some extent on such poorly known factors as the interatomic potential, ionization cross-sections and quantum-mechanical corrections to a treatment based on classical trajectories.

Once the theory of the SIMS process is properly understood, this technique should be capable of analyzing some aspects of the detailed surface geometry, such as the registries of adsorbates on a substrate and whether molecules adsorb associatively or dissociatively.

7.4 ISS (Ion-Scattering Spectroscopy)

In ISS, ions such as H^+, He^+ and Ar^+ are scattered off a surface and their energy distribution is observed[29)]. During the scattering process, the ions lose energy to the surface atoms. The collision process is usually so rapid (with kinetic energies of the order of 1 keV to 1 MeV) that a binary collision model is a good description of the situation. It is then easy to relate the energy loss ΔE to the mass M of the surface atoms involved:

$$\frac{E - \Delta E}{E} = \frac{1}{(1 + M/m)^2} \left\{ \cos\theta \pm \left[\left(\frac{M}{m} \right)^2 - \sin^2\theta \right]^{1/2} \right\}^2 ,$$

where E is the incident energy, m the ion mass and θ the laboratory scattering angle. Once the masses M of the surface atoms are known, the chemical composition follows.

Depending on energy and incidence direction (including channeling and blocking effects) the depth resolution of ISS can vary from a monolayer to about 300 Å. Quantitative evaluations of the chemical composition of surfaces may be achieved in special cases, but are generally hampered by uncertainties in the ion-atom scattering potential (especially at the lower energies), the possibility of multiple scatterings and the ever-present question of the depth distribution of individual species.

8 Methods Sensitive to Vibrational Structure of Surfaces

8.1 IR (Infrared Spectroscopy)

Absorption of infrared radiation by characteristic vibrations of a surface can be used to obtain information about that surface, by comparison with known absorption frequencies in molecules of known structure. Surface sensitivity is obtained by using small particles[30a)] and thin films[30b)] or, better, a multiple-reflection arrangement with optimized angles of incidence and reflection[30c)], in particular making work on single-crystal surfaces difficult: the difficulty comes from the very small cross-section for interaction (about 10^6 times smaller than for vibrational excitation by electrons). The accessible energy losses are somewhat restricted, so that only a limited set of vibrations can be detected. On the other hand, the resolution is largely sufficient for present-day needs. One great advantage of infrared spectroscopy is that it may be carried out while the surface is subjected to high gas pressures or in the presence of liquids: electron spectroscopies cannot be used under such circumstances. Electromagnetic laws, however, permit only vibration modes with significant components perpendicular to the surface to be detected. The knowledge of vibration frequencies has for example enabled statements to be made about the adsorption geometry of CO molecules on metal surfaces, supporting the view that such molecules bond by their C end to the surface, sometimes to single metal atoms (top bonding), sometimes to pairs of metal atoms (bridge bonding) or else to more metal atoms (hollow bonding). Such results are based on the knowledge of the vibration frequencies and geometries of metal carbonyl clusters: the metal–carbon bond vibrations can be recognized on the surface, while there appear no metal-oxygen vibration frequencies. It is furthermore found that these vibration frequencies can shift as a result of the coadsorption of electron-donor or electron-acceptor species. The change in electron density at the top metal layer presumably modifies the bonding between the metal and the carbon, and thereby also affects the C–O bond, shifting both frequencies.

However, the limitations inherent to IR spectroscopy have prevented it from being applied usefully to many other adsorbates.

8.2 HREELS (High-Resolution Electron Energy Loss Spectroscopy)

Electrons scattering off surfaces can lose energy in various ways. One of these involves excitation of the vibrational modes of atoms and molecules on the surface. The technique to detect vibrational excitation from surfaces by incident electrons is called high resolution electron energy loss spectroscopy (HREELS)[31]. It requires a beam of electrons with energy of incidence of about 10–20 eV, and with energy spread of at most about ±10 meV (that corresponds to about ±80 wave numbers). This energy spread is about an order of magnitude smaller than the energy spread of the electrons used presently for low-energy electron diffraction. This highly monochromatic electron beam, upon incidence on the surface, excites the vibrational mode of the different chemical bonds (M–H, M–C, M–O, C–C, C–H, etc. where M is a substrate atom). These modes have frequencies in the range of 100 to 300 meV (800 to 2400 wave numbers) and thus are readily detectable with the high energy resolution of this instrument. The electrons are back-reflected from the surface with energies equal to $E_{reflected} = E_{incident} - E_{vibration}$ and are detected by a suitable energy analyzer. The high resolution electron energy loss spectrum from hydrogen and deuterium adsorbed on the tungsten(110) crystal face is shown in Fig. 4.8. Not only is hydrogen readily detectable at coverages much less than a monolayer, but the isotope shifts on account of the different masses of H and D are also observed. The peaks are narrow and of high intensity and yield information on the location and structural symmetry of the sites where the surface atoms molecules are located.

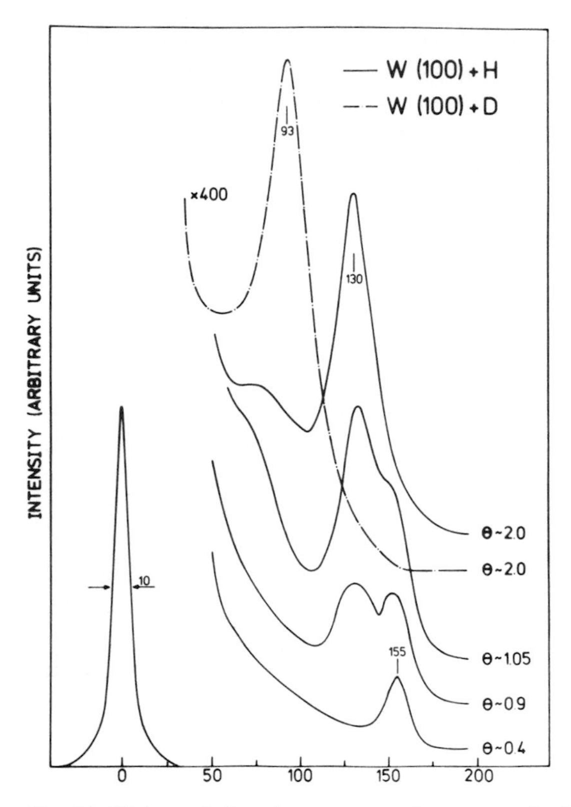

Fig. 4.8. High-resolution electron energy loss spectra for H and D adsorbed atomically on W(100). The elastic peak is shown at left. The loss energy for hydrogen is plotted along the horizontal axis. The coverage varies from $\theta = 0.4$ to $\theta = 2.0$ (saturation), exhibiting a change in adsorption site, while the deuterium spectrum is shown at $\theta = 2.0$ only. [After H. Froitzheim, H. Ibach and S. Lehwald, Phys. Rev. Lett. *36*, 1549 (1976).]

Adsorbed species with chemical bonds which are perpendicular to the surface plane are more readily detectable than chemical bonds that are parallel to the surface. The sensitivity of this technique appears to be 0.1 to 1% of a monolayer. Thus, the structure of the different adsorption sites that are filled up successively with increasing coverage of the adsorbate, can all be detected. There is a wealth of surface structural information that is becoming available from the high resolution electron loss spectra of adsorbed hydrocarbons and other molecules. A unique feature of this technique is that it is able to detect adsorbed hydrogen on the surface, because of the high-frequency vibrational modes of this atom when it is bound to the surface or to other adsorbed atoms (C–H, O–H, etc.). Hydrogen cannot be readily detected by other techniques such as LEED or other electron spectroscopies because of its small elastic and inelastic cross sections for scattering. This unique feature makes high-resolution electron loss spectroscopy an important tool for studying the surface chemistry of hydrogen, hydrocarbons and other hydrogen-containing molecules.

9 Simulation Methods

9.1 Model Calculations

Cutting across the domains of the various techniques mentioned above, are the model calculations[32]. These are theoretical attempts to predict the structure of surfaces from first principles. The model calculations differ from the theories mentioned in conjunction with the experimental techniques listed above, in that the former are not primarily designed to describe the interaction of a probe with a surface, although obviously much overlap exists. Thus the calculation of electronic states at surfaces seeks to describe from first principles a situation (the structure of the surface) that is analyzed experimentally by any of the techniques mentioned above, using external probes; but some of these techniques also involve the motion of electrons through the surface region: this motion in turn is clearly related to the electronic structure of the surface, and so the first-principles calculation and the surface-analysis technique may have and often do have much in common.

Model calculations are used in surface crystallography by comparing their predictions with observed values of electronic level energies, including densities of states (obtained mainly from UPS, XPS, and INS), of atomic and molecular binding energies (mainly from TDS and LEED), of vibration frequencies (from IR and EELS), of work function changes, etc. Model calculations can be subdivided into those that take a semi-infinite or film-like model of the surface and those that represent the surface by a cluster containing a finite number of atoms. The former are based on methods of solid-state physics, while the latter originate in molecular physics. The surface, especially when atomic adsorbates are present, is an intermediate situation between the crystal interior (with its three-dimensional periodicities) and molecules (with their limited dimensions). The loss of some translational symmetry is troublesome for solid-state theories, while the essentially infinite size of surfaces is troublesome for molecular physics. In a first stage therefore, model calculations were applied to idealized simple surfaces that could not be directly compared to real surfaces. However, progress has led to a situation in which now more and more cases of agreement with results from the other techniques of surface crystallography are reported. This will be seen in individual cases in Sect. VI.

9.2 Clusters

Recently it has been recognized that similarities exist between, on the one hand, metal surfaces with adsorption and, on the other hand, clusters consisting of a core of a few (1 to about 12) metal atoms, surrounded by various ligands, especially small attached molecules[33]. This opens up a new avenue for surface crystallography, since knowledge which is sometimes more easily gained on clusters may possibly be extrapolated to surfaces. Clusters are the nearest approximation to a crystal surface that is presently available experimentally for comparison. Such clusters can be analyzed structurally and energetically by various established methods, such as X-ray crystallography of clusters regularly arranged in single crystals. To what extent the analogy between clusters and surfaces is sufficiently close to make extrapolations from the former to the latter (or vice versa) is not clear yet, because of a paucity of directly comparable cases in the experiments performed to date.

One promising class of structures concerns CO adsorption on metals and its metal carbonyl cluster counterparts. In such clusters, CO is found to often attach itself molecularly to the metal frame, with the C end bonded to either one, two or three metal atoms (terminal, edge and face bonding, respectively). There are corresponding clear differences in bond lengths (not only in the metal–C bond but also in the CO bond) and in binding energies between the different bonding configurations. There are also clusters containing hydrocarbon ligands, which are useful analogies in the study of hydrocarbon adsorption. For example, the adsorption of ethylene (C_2H_4) and acetylene (C_2H_2) may be studied by comparison with the clusters $Co_3(CCH_3)(CO)_9$, $Ru_3(CCH_3)$-$H_3(CO)_9$, Br_3CCH_3, $(Ph_3P)_2Pt(C_2Ph_2)$, $Co_2(C_2Ph_2)(CO)_6$, $Rh_2(h^5$-$C_5H_5)_2(CO)_2$-$(CF_3C_2CF_3)$ and $Os_3(C_2Ph_2)(CO)_{10}$.

References

1. For more extensive general reviews, see for example:
   a. Somorjai, G. A.: Principles of Surface Chemistry, Englewood Cliffs, N.J.: Prentice-Hall 1972;
   b. Ertl, G., Küppers, J.: Low energy electrons and surface chemistry, Weinheim: Verlag Chemie 1974
2. See also:
   a. Pendry, J. B.: Low energy electron diffraction, London: Academic Press 1974;
   b. Strozier, J. A. Jr., Jepsen, D. W., Jona, F.: In: Surface Physics of Materials, ed. J. M. Blakeley, New York: Academic Press 1975;
   c. Kesmodel, L. L., Somorjai, G. A.: Acc. Chem. Res., *9*, 392 (1976);
   d. various articles in "Characterization of Metal and Polymer Surfaces, (ed. L.-H. Lee), New York: Academic Press 1977, Vol. 1, Part III;
   e. Jona, F.: Surf. Sci. *68*, 204 (1977);
   f. Van Hove, M. A., Tong, S. Y.: Surface crystallography by low-energy electron diffraction: theory, computation and structural results, Heidelberg: Springer 1979
3. a. Henzler, M.: Surf. Sci. *19*, 159 (1970);
   b. Laramore, G. E., Houston, J. E., Park, R. L.: J. Vac. Sci. Technol. *10*, 196 (1973)
4. a. Houston, J. E., Park, R. L.: Surf. Sci. *21*, 209 (1970);
   b. Ertl, G., Küppers, J.: Surf. Sci. *21,* 61, (1970);
   c. Jagodzinski, H., Wolf, D., Moritz, W.: Surf. Sci., *77*, 223, 249, 265 and 283 (1978)
5. Ertl, G., Schillinger, D.: J. Chem. Phys. *66*, 2569 (1977)
6. Wang, G.-C., Lu, T.-M., Lagally, M. G.: J. Chem. Phys. *69*, 479 (1978)
7. Behm, R. J., Christmann, K., Ertl, G.: Sol. St. Commun. *25*, 763 (1978)
8. a. Lagally, M., Ngoc, T. C., Webb, M. B.: Phys. Rev. Lett. *26,* 1557 (1971);
   b. Tucker, C. W., Duke, C. B.: Surf. Sci. *29*, 237 (1972)
9. a. Adams, D. L., Landman, U.: Phys. Rev. Lett. *33*, 585 (1974);
   b. Cohen, P. I., Unguris, J., Webb, M. B.: Surf. Sci. *58*, 429 (1976);
   c. Adams, D. L., Landman, U.: Phys. Rev. *B15*, 3775 (1977);
   d. Cunningham, S. L., Chan, C.-M., Weinberg, W. H.; Phys. Rev. *B18*, 1537 (1978)
10. Tong, S. Y., Van Hove, M. A., Mrstik, B. J.: Proc. 7th Intern. Vac. Congr. and 3rd Intern. Confer. on Solid Surfaces, Vienna, p. 2407 (1977)

11. a. Zimmer, R. S., Holland, B. W.: J. Phys. *C8*, 2395 (1975);
b. Tong, S. Y., Van Hove, M. A.: Phys. Rev. *B16*, 1459 (1977)
12. Feder, R.: Phys. Stat. Sol. (b) *62*, 135 (1974)
13. Duke, C. B., Liebsch, A.: Phys. Rev. *B9*, 1126 and 1150 (1974)
14. a. Menadue, J. F.: Acta Cryst. *A28*, 1 (1972);
b. Masud, N., Pendry, J. B.: J. Phys. *C9*, 1833 (1976);
c. Masud, N., Kinniburgh, C. G., Pendry, J. B.: J. Phys. *C10*, 1 (1977)
15. a. Crewe, A. V., Wall, J., Langmore, J.: Science *168*, 133 (1970);
b. Cowley, J. M.: Diffraction Physics, Amsterdam: North-Holland 1975;
c. Isaacson, M. S., Langmore, J., Parker, N. W., Kopf, D., Utlaut, M.: Ultramicroscopy *1*, 359 (1976);
d. Yagi, K., Takayanagi, K., Kobayashi, K., Osakabe, N., Tanishiro, Y., Honjo, G.: Electron Microscopy, 1978, Vol. I (Proc. 9th Intern. Congr. on El. Micr., Toronto, 1978), J. M. Sturgess, Ed., Microscopical Society of Canada, Toronto (1978)
16. a. Müller, E. W.: Z. Physik *136*, 131 (1951);
b. Müller, E. W.: J. Appl. Phys. *27*, 474 (1956);
c. Müller, E. W.: In: Advances in Electronics and Electron Physics, Vol. XIII, L. Marton, ed. New York: Academic Press 1960;
d. Müller, E. W., Tsong, T. T.: Field Ion Microscopy, New York: American Elsevier 1969;
e. Ehrlich, G.: Surf. Sci. *63*, 422 (1977);
f. Ayrault, G., Ehrlich, G.: J. Chem. Phys. *60*, 281 (1974)
17. a. Heiland, W., Taglauer, E.: Surf. Sci. *68*, 96 (1977);
b. Saris, F. W., Van der Veen, J. F.: Proc. 7th Intern. Vac. Congr. and 3rd Intern. Confer. on Solid Surfaces, Vienna, p. 2503 (1977);
c. Feldman, L. C., Kauffman, R. L., Silverman, P. J., Zuhr, R. A., Barrett, J. H.: Phys. Rev. Lett. *39*, 38 (1977);
d. Van der Veen, J. F., Tromp, R. M., Smeenk, R. G., Saris, F. W.: Surf. Sci. *82*, 468 (1979)
18. a. Masel, R. I.: Merrill, R. P., Miller, W. H.: J. Vac. Sci. Technol. *13*, 355 (1976);
b. Beeby, J. L.: J. Phys. *C7*, 1 (1975)
c. Ceyer, S. T., Gale, R. J., Bernasek, S. L., Somorjai, G. A.: J. Chem. Phys. *64*, 1934 (1976);
d. Ceyer, S. T., Somorjai, G. A.: Am. Rev. Phys. Chem. *28*, 477 (1977)
19. a Lee, P. A.: Phys. Rev. *B13*, 5261 (1976);
b. Citrin, P. H., Eisenberger, P., Hewitt, R. C.: J. Vac. Sci. Technol. *15*, 449 (1978)
20. a. Eastman, D. E., Cashion, J. K.: Phys. Rev. Lett. *27*, 1520 (1971);
b. Bradshaw, A. M., Cederbaum, L. S., Domcke, W.: in Structure and Bonding, Vol. 24 (1975);
c. Plummer, E. W.: In: Photoemission and the electronic properties of surfaces, B. Feuerbacher, B. Fitton and R. F. Willis, eds., London: Wiley 1978
21. Gadzuk, J. W.: Phys. Rev. *B14*, 2267 (1976)
22. a. Gadzuk, J. W.: Phys. Rev. *B10*, 5030 (1974);
b. Pendry, J. B.: Surf. Sci. *57*, 679 (1976);
c. Liebsch, A.: Phys. Rev. Lett. *38*, 248 (1977);
d. Jacobi, K., Scheffler, M., Kambe, K., Forstmann, F.: Sol. St. Comm. *22*, 17 (1977);
e. Li, C. H., Tong, S. Y.: Phys. Rev. Lett. *40*, 46 (1978)
23. a. Hagstrum, H. D.: Phys. Rev. *150*, 495 (1966);
b. Becker, G. E., Hagstrum, H. D.: J. Vac. Sci. Technol. *10*, 31 (1973);
c. Hagstrum, H. D., Becker, G. E.: J. Vac. Sci. Technol. *14*, 369 (1977)
24. a. Johnson, P. D., Delchar, T. A.: Surf. Sci. *77*, 400 (1978)
b. Wang, S. W., Ertl, G.: to be published;
c. Conrad, H., Ertl, G., Küppers, J., Wang, S. W., Gérard, K., Haberland, H.: Phys. Rev. Lett. *42*, 1082 (1979)
25. a. Herring, C., Nichols, M. H.: Rev. Mod. Phys. *21*, 185 (1949);
b. Lang, N. D., Kohn, W.: Phys. Rev. *B3*, 1215 (1971);
c. Topping, J.: Proc. Roy. Soc. (London), *A114*, 67 (1927)
26. a. Palmberg, P. W.: In: Electron spectroscopy, D. A. Shirley, ed., Amsterdam: North Holland 1972;

b. Carlson, T. A.: Photoelectron and Auger Spectroscopy, New York: Plenum Press 1975
27. a. Redhead, P. A.: Vacuum *12*, 203 (1962);
b. Petermann, L. A.: Progress in Surf. Science, Vol. 3, S. G. Davison, ed., Oxford-New York: Pergamon Press 1972;
c. King, D. A.: Surf. Sci. *47*, 384 (1975)
28. a. Benninghoven, A.: Surf. Sci. *35*, 427 (1973);
b. Garrison, B. J., Winograd, N., Harrison, D. E. Jr.: Phys. Rev. *B18*, 6000 (1978)
29. a. Heiland, W., Taglauer, E.: Surf. Sci. *68*, 96 (1977);
b. Saris, F. W., Van der Veen, J. F.: Proc. 7th Intern. Vac. Congr. & 3rd Intern. Conf. Solid Surfaces, Vienna (1977), p. 2503
30. a. Hair, M. L.: Infrared spectroscopy in surface chemistry, New York: Dekker 1967;
b. Bradshaw, A. M., Pritchard, J.: Surf. Sci. *17*, 372 (1969);
c. Tompkins, H. G., Greenler, R. G.: Surf. Sci. *28*, 194 (1971)
31. a. Propst, F. M., Piper, T. C.: J. Vac. Sci. Technol. *4*, 53 (1967);
b. Ibach, H.: Phys. Rev. Lett. *24*, 1416 (1970);
c. Andersson, S.: Sol. St. Comm. *21*, 75 (1977)
32. a. Overviews of theoretical methods in surface model calculations are given in two chapters by T. B. Grimley and R. P. Messmer, resp., In: The nature of the surface chemical bond, T. N. Rhodin and G. Ertl, eds., Amsterdam: North Holland 1978;
see also:
b. Einstein, T. L., Schrieffer, J. R.: Phys. Rev. *B7*, 3629 (1973);
c. Grimley, T. B., Pisani, C.: J. Phys. *C7*, 2831 (1974);
d. Lang, N. D., Williams, A. R.: Phys. Rev. Lett. *34*, 531 (1975);
e. Gunnarson, O., Hjelmberg, H., Lundqvist, B. I.: Phys. Rev. Lett. *37*, 292 (1976);
f. Wang, S. W., Weinberg, W. H.: Surf. Sci. *77*, 14 and 29 (1978);
g. Fassaert, D. J. M., Van der Avoird, A.: Surf. Sci. *55*, 291 and 313 (1976);
h. Louie, S. G., Ho, K. M., Chelikowsky, J. R., Cohen, M. L.: Phys. Rev. Lett. *37*, 1289 (1976);
i. Kasowski, R. V.: Phys. Rev. Lett. *37*, 219 (1976);
j. Appelbaum, J. A., Baraff, G. A., Hamann, D. R.: Phys. Rev. *B14*, 588 (1976);
k. Harrison, W. A.: Surf. Sci. *55*, 1 (1976);
l. Bullett, D. W., Cohen, M. L.: J. Phys. Chem. *10*, 2083 and 2101 (1977);
m. Messmer, R. P., Knudsen, S. K., Johnson, K. H., Diamond, J. B., Yang, C. Y.: Phys. Rev. *B13*, 1396 (1976);
n. Batra, I. P., Robaux, O.: Surf. Sci. *49*, 653 (1975);
o. Marshall, R. F., Blint, R. J., Kunz, A. B.: Phys. Rev. *B13*, 3333 (1976);
p. Walch, S. P., Goddard III, W. A.: J. Am. Chem. Soc. *98*, 7908 (1976);
q. Anderson, A. B.: J. Chem. Phys. *66*, 2173 (1977)
33. a. Chini, P., Longoni, G., Albano, V. G.: Adv. Organomet. Chem. *14*, 285 (1976);
b. Muetterties, E. L., Rhodin, T. N., Band, E., Brucker, C. F., Pretzer, W. R.: Chem. Revs., to be published (1979)

V Overview of the Ordering of Adsorbates

1 Introduction

While only about 40 surface structures have been analyzed by methods of surface crystallography in order to determine the precise location of adsorbed atoms, or molecules, nearly 1000 ordered surface structures of adsorbates have been reported. It appears that almost any adsorbate monolayer may be ordered to form at least one and frequently several structures under appropriate conditions of gas exposure and temperature. The proper experimental conditions achieve a balance among the various surface forces (heats of adsorption, activation energies of surface and bulk diffusion, etc.) that facilitate the formation of large ordered domains that yield sharp diffraction features. The ordering of adsorbed monolayers is a very sensitive function of temperature. For example, the lowering of the temperature of rhodium single crystals from 300 to 270 K greatly increases the size of the ordered domains of CO, O_2 and other adsorbates which, in turn, visibly improves the quality of the diffraction patterns. Similar observations are reported commonly for other adsorbed monolayer systems as well. The adsorbate ordering obviously also depends strongly on coverage, since a particular periodic arrangement of adsorbates at one coverage cannot freely accommodate a change in coverage. Among other similar examples, Pb deposited on Au(100) produces $c(2 \times 2)$, $c(7\sqrt{2} \times \sqrt{2})R45°$, $c(3\sqrt{2} \times \sqrt{2})R45°$ and $c(6 \times 2)$ arrangements as the coverage is varied. (The unit cell notation is discussed in Sect. III.)

In this section we present comprehensive tabulations of the observed ordered structures for any adsorbate on any substrate. For most of these cases the surface structure has not been analyzed beyond the implications of the unit cell shape, size and orientation. Many of these structures are good candidates for a structural analysis of the binding sites, bond lengths and bond angles. It is hoped that the list of geometrically analyzed structures (discussed in detail in Sects. VI and VII) will grow rapidly so as to present an expanding base for the extraction of fundamental laws governing the adsorption phenomenon.

Low Miller index surfaces of metallic single crystals are the most commonly used substrates in LEED investigations. The reasons for their widespread use are that they have the lowest surface free energy and therefore are the most stable, have the highest rotational symmetry and are the most densely packed. Also, in the case of transition metals and semiconductors they are chemically less reactive than the higher Miller index crystal faces.

The metal substrates used in the LEED experiments have either face centered cubic (fcc), body centered cubic (bcc) or hexagonal closed packed (hcp) crystal structures. For the cubic metals the (111), (100) and (110) planes are the low Miller index surfaces and they have threefold, fourfold and twofold rotational symmetry, respectively. The top layer of a (111) surface actually has sixfold symmetry, but the rotational symmetry of the top layers together is threefold. Since the near surface region can influence where gases adsorb on the surface and the LEED intensities exhibit threefold rotational symmetry at normal incidence, the (111) surface will be considered to have threefold rotational symmetry. Although most of the adsorption studies have been carried out on fcc and bcc crystals, there have been several studies reported on hcp crystals. For hcp metals the basal or (0001) plane is the surface most frequently studied by LEED investigations and it is the most densely packed plane having threefold rotational symmetry.

2 Metals on Metals

In Table 5.1 the surface structures of ordered metal monolayers adsorbed on metal surfaces are listed. For each substrate, the crystallographic structure, the distance between nearest neighbors, and the heat of sublimation (that is proportional to the surface free energy) are given. For each metal adsorbate the identical information is provided along with the technique of deposition and all the ordered surface structures that form with increasing coverage.

One of the striking results of these studies that is revealed by the inspection of Table 5.1 is the predominance of the formation of ordered monolayers regardless of the relative magnitudes of the surface free energies of the substrate and adsorbate metals. Surface thermodynamic considerations would predict monolayer formation only when the total surface free energy is minimized this way (i.e., during the deposition of a metal of lower surface free energy on a metal substrate of higher surface free energy). If these circumstances are not met the growth of three-dimensional crystallites is predicted (i.e., when the adsorbate surface free energy is greater than that of the substrate) to minimize the total surface energy. However, the experimental data indicate that regardless of the surface free energy differences (for example, even for Mo on Ni, Pt on Au and Cu on Zn), ordered monolayer deposits form.

There is one exception to the formation of ordered monolayers: Fe on W forms three-dimensional crystallites even though surface thermodynamic considerations would predict monolayer formation.

At low adsorbate coverages the surface structure of the deposited metal is determined by the substrate periodicity. Thus, under these conditions the adsorbate-substrate interaction is predominant. At higher coverages the adsorbate may continue to follow the substrate periodicity or form coincidence structures with new periodicities that are unrelated to the substrate periodicity. The ordering geometry of large-radius metallic adatoms (especially K, Rb and Cs) shows relatively little dependence on the substrate lattice: they tend to form hexagonal close-packed layers on any metal substrate. It appears that for these systems the adsorbate-adsorbate interaction predominates during ordering.

The available data are inadequate to permit a detailed analysis of the various factors that control the ordering of metal monolayers on metal surfaces. It is likely that both the electronic interaction between the two metals and the relative atomic sizes should be important in determining the nature of ordering in the monolayer.

3 Non-Metallic Adsorbates

In Tables 5.2 to 5.7 we list the observed adsorbate surface structures (excluding metallic adsorbates, listed in Table 5.1). The substrates are classified according to the rotational symmetry of their surfaces: three-fold in Table 5.2, four-fold in Table 5.3, and two-fold in Table 5.4. Stepped surfaces are considered in Table 5.6; this is preceded by Table 5.5 which lists the relation between the special notation for stepped surfaces and the conventional surface plane notation. Finally, structures formed with organic adsorbates are brought together in Table 5.7. Most of the substrates in Tables 5.2, 5.3 and 5.4 are low index faces and the gases adsorbed are, for the most part, small inorganic molecules such as H_2, O_2, N_2, CO and NO. Inspection of the tables permits one to propose two general rules that are usually obeyed during the adsorption of these small molecules: (1) the observed surface structures have the same rotational symmetry as the substrate, and (2) the unit cell of the surface structure is the smallest allowed by the molecular dimensions and adsorbate-adsorbate interactions.

The frequent occurrence of ordered fractional-coverage adsorption indicates that adsorbate-adsorbate interactions at close range ($\lesssim 5$ Å) are often repulsive. Island formation can occur simultaneously, showing that at larger separations these interactions can become attractive.

There is also a general tendency for adsorbates of any type to form identical superstructures on different substrates of a given symmetry, showing the effect of adsorbate-adsorbate interactions. For example, oxygen forms a (2×2) superstructure on the hexagonal faces of Ag, Cu, Ir, Nb, Ni, Pd, Pt, Re, Rh, and Ru. This is most obvious for the physisorption of rare gases, where the adsorbate-substrate interactions parallel to the surface are so small that a hexagonal close-packed layer is formed even on substrates of different surface symmetry and greater roughness, such as with Xe on Cu(100), Cu(110), Cu(211) and Cu(311). This hexagonal overlayer has been analyzed for Xe on Ag(111) and found to correspond to the (111) plane of the fcc inert gas solid (cf. Sect. VI).

In the last few years LEED studies of high Miller index or stepped surfaces have become more frequent. Almost all of these studies have been on fcc metals, where the atomic structure of these surfaces consists of periodic arrays of terraces and steps. A nomenclature which is more descriptive of the actual surface configuration has been developed for these surfaces, as described in Section III. In Table 5.5 the stepped surface nomenclature for several high Miller index surfaces of fcc crystals has been tabulated. In Fig. 5.1 the location of these high Miller index surfaces are shown on the

Table 5.1. Adsorption properties of metal monolayers on metal substrates. The clean substrate properties are also given for comparison. Substrates are ordered by lattice type (fcc, bcc, hcp, cubic, diamond and rhombic). The structures, nearest neighbor distances and heats of vaporization refer to the bulk material of the substrate or the adsorbate. VD, ID and S stand for vapor deposition, ion beam deposition and surface segregation, respectively. TD, WF and TED stand for thermal desorption, work function measurements and transmission electron diffraction, respectively

Substrate metal	Adsorbed metal	Structure	Nearest neighbor distance	Heat of vaporization (kcal/g atom)	Deposition technique	Substrate orientation	Technique of investigation	Surface structures observed	References
Rh	–	fcc	2.69	127	–	–	–	–	–
Rh	Fe	bcc	2.48	85	VD	(100)	TED	Fe(100) and Fe(110) ‖ Rh(100)	13)
Ir	–	fcc	2.71	160	–	–	–	–	–
Ir	Cr	bcc	2.50	73	VD	(111)	LEED-AES-WF	hexagonal	14)
Ir	Au	fcc	2.88	82	VD	(111)	LEED-AES-WF	$\begin{pmatrix}1 & 0\\0 & 1\end{pmatrix}$	14)
Ni	–	fcc	2.49	91	–	–	–	–	–
Ni	Na	bcc	3.66	24	ID/VD	(100)	LEED-WF	$\begin{pmatrix}1 & 1\\1 & \bar{1}\end{pmatrix}$	15, 13, 16-20)
					ID	(111)	LEED	hexagonal	16, 19, 21)
					ID	(110)	LEED	disordered structures, hexagonal	16, 19, 21)
Ni	K	bcc	4.52	19	ID/VD	(100)	LEED-WF/LEED	$\begin{pmatrix}4 & 0\\0 & 2\end{pmatrix}$/hexagonal	15, 18, 22)
					ID	(110)	LEED	disordered structures	16)
Ni	Cs	bcc	5.23	16	ID	(100)	LEED-WF/LEED	$\begin{pmatrix}2 & 0\\0 & 2\end{pmatrix}$ / hexagonal	15, 23)
					ID	(110)	LEED	disordered structures	16)
Ni	Ba	bcc	4.35	36	VD	(100)	LEED-WF	disordered overlayer	15)
Ni	Cr	bcc	2.50	73	VD	(100)	TED	$\begin{pmatrix}1 & 0\\0 & 1\end{pmatrix}$	23)
Ni	Mo	bcc	2.72	128	S	(111)	LEED	$\begin{pmatrix}5 & 0\\0 & 5\end{pmatrix}, \begin{pmatrix}4 & 0\\0 & 4\end{pmatrix}, \begin{pmatrix}2 & 0\\5 & 10\end{pmatrix}, \begin{pmatrix}1 & 0\\5 & 10\end{pmatrix}$	8)
Ni	Fe	bcc	2.48	85	VD	(100)	TED	(110)Fe ‖ (100) Ni	13)
Ni	Co	hcp	2.50	93	VD	(100)	TED	$\begin{pmatrix}1 & 0\\0 & 1\end{pmatrix}$	25)
Ni	Cu	fcc	2.56	73	VD	(100)	RHEED-AES	$\begin{pmatrix}1 & 0\\0 & 1\end{pmatrix}$	26)

Ni	Ag	fcc	2.89	61	VD	(111)	LEED/AES	$\begin{pmatrix}6&0\\0&6\end{pmatrix}$	27, 28)
Ni	Au	fcc	2.88	82	S	(111)	LEED-AES/ LEED	$\begin{pmatrix}6&0\\0&6\end{pmatrix}$, $\begin{pmatrix}13&0\\0&13\end{pmatrix}$/ $\begin{pmatrix}6&0\\0&6\end{pmatrix}$	29, 30)
Ni	Pb	fcc	3.50	42	VD	(100)	LEED	$\begin{pmatrix}1&1\\1&\bar{1}\end{pmatrix}$, $\begin{pmatrix}1&1\\\bar{2}&3\end{pmatrix}$	31)
					VD	(111)	LEED	$\begin{pmatrix}1&1\\\bar{1}&2\end{pmatrix}$, $\begin{pmatrix}7&0\\0&7\end{pmatrix}$, $\begin{pmatrix}13&0\\0&13\end{pmatrix}$, $\begin{pmatrix}3&0\\0&3\end{pmatrix}$ / hexagonal rotated ±3°, $\begin{pmatrix}3&0\\0&3\end{pmatrix}$, $\begin{pmatrix}1&1\\\bar{1}&2\end{pmatrix}$	31, 32)
					VD	(110)	LEED	$\begin{pmatrix}1&1\\1&\bar{1}\end{pmatrix}$, $\begin{pmatrix}3&0\\0&1\end{pmatrix}$, $\begin{pmatrix}4&0\\0&1\end{pmatrix}$, $\begin{pmatrix}5&0\\0&1\end{pmatrix}$	31)
Pd	–	fcc	2.75	90	–	–	–	–	–
Pd	Fe	bcc	2.48	85	VD	(100)	TED	(100)Fe ‖ (100)Pd and (110)Fe ‖ (100)Pd	13)
Pd	Ni	fcc	2.49	91	VD	(100)	TED	$\begin{pmatrix}1&0\\0&1\end{pmatrix}$	33)
Pd	Ag	fcc	2.89	61	VD	(100)	LEED	$\begin{pmatrix}1&0\\0&1\end{pmatrix}$	34)
Pd	Au	fcc	2.88	82	VD	(100)	LEED/TED	$\begin{pmatrix}1&0\\0&1\end{pmatrix}$	34, 35, 36)
					VD	(111)	TED	$\begin{pmatrix}1&0\\0&1\end{pmatrix}$	36)
Cu	–	fcc	2.56	73	–	–	–	–	–
Cu	Fe	bcc	2.48	85	VD	(100)	TED	$\begin{pmatrix}1&0\\0&1\end{pmatrix}$	13, 37, 38)
					VD	(111)	LEED-AES	$\begin{pmatrix}1&0\\0&1\end{pmatrix}$	39)
Cu	Co	hcp	2.50	93	VD	(100)	TED	$\begin{pmatrix}1&0\\0&1\end{pmatrix}$	40, 41)
Cu	Ni	fcc	2.49	91	VD	(100)	TED	$\begin{pmatrix}1&0\\0&1\end{pmatrix}$	42)
					VD	(111)	LEED/RHEED	$\begin{pmatrix}1&0\\0&1\end{pmatrix}$	43, 44, 45)

Table 5.1 (continued)

Substrate metal	Adsorbed metal	Structure	Nearest neighbor distance	Heat of vaporization (kcal/g atom)	Deposition technique	Substrate orientation	Technique of investigation	Surface structures observed	References
Cu	Ag	fcc	2.89	61	VD	(100)	LEED	$\begin{pmatrix} 2 & 0 \\ \bar{1} & 5 \end{pmatrix}$	34, 36)
					VD	(111)	LEED/ RHEED,TED	$\begin{pmatrix} 8 & 0 \\ 0 & 8 \end{pmatrix}$ / three dimensional crystals	46, 45, 47-50)
Cu	Au	fcc	2.88	82	VD	(100)	LEED	$\begin{pmatrix} 1 & 1 \\ 1 & \bar{1} \end{pmatrix}$, $\begin{pmatrix} 2 & 0 \\ \bar{1} & 7 \end{pmatrix}$	34, 51)
					VD	(111)	LEED-AES/ RHEED	$\begin{pmatrix} 2/3 & 2/3 \\ \overline{2/3} & 4/3 \end{pmatrix}$, $\begin{pmatrix} 2 & 0 \\ 0 & 2 \end{pmatrix}$ / three dim. crystals	52, 47, 53)
					VD	(110)	LEED-AES	$\begin{pmatrix} 1 & 0 \\ \overline{1/2} & 3/2 \end{pmatrix}$, $\begin{pmatrix} 1 & 0 \\ 0 & 2 \end{pmatrix}$, $\begin{pmatrix} 2 & 0 \\ 0 & 2 \end{pmatrix}$, complex structures	52)
Cu	Sn	diam	2.81	70	S	(100)	LEED-AES	$\begin{pmatrix} 2 & 0 \\ 0 & 2 \end{pmatrix}$	54)
					S	(111)	LEED-AES	$\begin{pmatrix} 1 & 1 \\ \bar{1} & 2 \end{pmatrix}$	54)
Cu	Pb	fcc	3.50	42	VD	(100)	LEED/LEED-AES-TD-TED	$\begin{pmatrix} 2 & 2 \\ \bar{2} & 2 \end{pmatrix}$, $\begin{pmatrix} 1 & 1 \\ \bar{2} & 3 \end{pmatrix}$	55, 56-58, 59)
					VD	(111)	LEED/LEED-AES-TD	$\begin{pmatrix} 4 & 0 \\ 0 & 4 \end{pmatrix}$	55, 58)
					VD	(110)	LEED/LEED-AES	$\begin{pmatrix} 1 & 1 \\ \bar{1} & 1 \end{pmatrix}$, $\begin{pmatrix} 5 & 0 \\ 0 & 1 \end{pmatrix}$/$\begin{pmatrix} 1 & 1 \\ \bar{1} & 1 \end{pmatrix}$, $\begin{pmatrix} 4 & 0 \\ 0 & 1 \end{pmatrix}$, $\begin{pmatrix} 5 & 0 \\ 0 & 1 \end{pmatrix}$	55, 56)
					VD	(711)	LEED-AES/ LEED-AES-TD	$\begin{pmatrix} 4 & 0 \\ 0 & 1 \end{pmatrix}$	56, 58)
					VD	(511)	LEED-AES	$\begin{pmatrix} 4 & 0 \\ 0 & 1 \end{pmatrix}$	56)
					VD	(311)	LEED-AES	$\begin{pmatrix} 3 & 1 \\ \bar{2} & 1 \end{pmatrix}$, $\begin{pmatrix} 4 & 0 \\ 0 & 2 \end{pmatrix}$	58)
					VD	(211)	LEED-AES	$\begin{pmatrix} 4 & 0 \\ 0 & 1 \end{pmatrix}$	58)

Cu	Bi	rhomb	3.07	43	VD	(100)	LEED/LEED-AES	$\begin{pmatrix}2 & 0\\0 & 2\end{pmatrix}, \begin{pmatrix}1 & \bar{1}\\1 & 1\end{pmatrix}, \begin{pmatrix}\bar{1} & 1\\4 & 5\end{pmatrix}, \begin{pmatrix}\bar{3} & 4\\4 & 5\end{pmatrix}$	[illegible]
					VD	(111)	LEED	$\begin{pmatrix}1 & 1\\\bar{1} & 2\end{pmatrix}, \begin{pmatrix}2 & \bar{1}\\0 & 2\end{pmatrix}, \begin{pmatrix}2 & 3\\1 & 2\end{pmatrix}$	61)
Ag	–	fcc	2.89	61	–	–	–	–	–
Ag	Na	bcc	3.66	24	VD	(111)	LEED-AES-TD	$\begin{pmatrix}1 & 0\\0 & 1\end{pmatrix}$	62)
					VD	(110)	LEED-AES-TD	$\begin{pmatrix}1 & 0\\0 & 1\end{pmatrix}$	63)
Ag	Rb	bcc	4.84	18	VD	(111)	LEED-AES-TD	$\begin{pmatrix}1 & 0\\0 & 1\end{pmatrix}$	64)
Ag	Mg	hcp	3.20	32	VD	(111)	TED	disordered overlayer	65)
Ag	Cr	bcc	2.50	73	VD	(111)	TED	disordered overlayer	65)
Ag	Co	hcp	2.50	93	VD	(111)	TED	disordered overlayer	65)
Ag	Ni	fcc	2.49	91	VD	(100)	TED	$\begin{pmatrix}1 & 0\\0 & 1\end{pmatrix}$	66)
					VD	(111)	TED/RHEED	hexagonal overlayer	65, 67)
Ag	Pd	fcc	2.75	90	VD	(111)	TED	disordered overlayer	65)
Ag	Cu	fcc	2.56	73	VD	(100)	TED	$\begin{pmatrix}1 & 0\\0 & 1\end{pmatrix}$	66)
					VD	(111)	RHEED/TED	hexagonal overlayer	68-70)
Ag	Au	fcc	2.88	82	VD	(100)	TED	$\begin{pmatrix}1 & 0\\0 & 1\end{pmatrix}$	35)
					VD	(111)	LEED-AES/TED	$\begin{pmatrix}1 & 0\\0 & 1\end{pmatrix}$	65, 71, 72)
Ag	Zn	hcp	2.66	27	VD	(111)	TED	no condensation	65)
Ag	Cd	hcp	2.98	24	VD	(111)	TED	no condensation	65)
Ag	Al	fcc	2.86	68	VD	(111)	TED	disordered overlayer	65)
Ag	Tl	hcp	3.46	39	VD	(111)	TED	hexagonal overlayer	65)
Ag	Sn	diam	2.81	70	VD	(111)	TED	disordered overlayer	65)
Ag	Pb	fcc	3.50	42	VD	(111)	TED	hexagonal overlayer	65, 73)
Ag	Sb	rhomb	2.91	47	VD	(111)	TED	disordered overlayer	65)
Ag	Bi	rhomb	3.07	43	VD	(111)	TED	disordered overlayer	65)
Au	–	fcc	2.88	82	–	–	–	–	–
Au	Na	bcc	3.66	24	VD	(100)	LEED	series of structures, hexagonal	74)
Au	Cr	bcc	2.50	73	VD	(111)	LEED-AES-WF	hexagonal	75)

Table 5.1 (continued)

Substrate metal	Adsorbed metal	Structure	Nearest neighbor distance	Heat of vaporization (kcal/g atom)	Deposition technique	Substrate orientation	Technique of investigation	Surface structures observed	References
Au	Fe	bcc	2.48	85	VD	(100)	TED	$\begin{pmatrix}1 & 0\\ 0 & 1\end{pmatrix}$	76-78)
					VD	(111)	TED	$\begin{pmatrix}1 & 0\\ 0 & 1\end{pmatrix}$	76, 78-80)
Au	Pd	fcc	2.75	90	VD	(100)	LEED/TED	$\begin{pmatrix}1 & 0\\ 0 & 1\end{pmatrix}$	34, 81)
					VD	(111)	TED	$\begin{pmatrix}1 & 0\\ 0 & 1\end{pmatrix}$	36, 82)
Au	Pt	fcc	2.77	122	VD	(100)	LEED-AES/ TED	$\begin{pmatrix}1 & 0\\ 0 & 1\end{pmatrix}$	3, 4)
					VD	(111)	TED	$\begin{pmatrix}1 & 0\\ 0 & 1\end{pmatrix}$	83)
Au	Cu	fcc	2.56	73	VD	(100)	LEED	$\begin{pmatrix}1 & 0\\ 0 & 1\end{pmatrix}$	34)
					VD	(111)	RHEED	extra lines	84, 85)
Au	Ag	fcc	2.89	61	VD	(100)	LEED/TED	$\begin{pmatrix}1 & 0\\ 0 & 1\end{pmatrix}$	34, 39, 86, 87)
						(111)	LEED-AES	$\begin{pmatrix}1 & 0\\ 0 & 1\end{pmatrix}$	71)
Au	Pb	fcc	3.50	42	VD	(100)	LEED-AES	$\begin{pmatrix}1 & 1\\ 1 & \bar{1}\end{pmatrix}, \begin{pmatrix}1 & 1\\ \bar{3} & 4\end{pmatrix}, \begin{pmatrix}1 & 1\\ \bar{1} & 2\end{pmatrix}, \begin{pmatrix}2 & 0\\ \bar{1} & 3\end{pmatrix}$	5, 6)
					VD	(111)	LEED-AES	hexagonal rotated ±5°	6, 88)
					VD	(110)	LEED-AES	$\begin{pmatrix}1 & 0\\ 0 & 3\end{pmatrix}, \begin{pmatrix}1 & 0\\ 0 & 1\end{pmatrix}, \begin{pmatrix}7 & 0\\ 0 & 1\end{pmatrix}, \begin{pmatrix}7 & 0\\ 0 & 3\end{pmatrix}, \begin{pmatrix}4 & 0\\ 0 & 4\end{pmatrix}$	6, 88)
					VD	(11,1,1)	LEED-AES	$\begin{pmatrix}1 & 1\\ 1 & \bar{1}\end{pmatrix}, \begin{pmatrix}2 & 0\\ \bar{1} & 3\end{pmatrix}$	6)
					VD	(911)	LEED-AES	$\begin{pmatrix}1 & 1\\ 1 & \bar{1}\end{pmatrix}, \begin{pmatrix}2 & 0\\ \bar{1} & 3\end{pmatrix}$	6)
					VD	(711)	LEED-AES	$\begin{pmatrix}1 & 1\\ 1 & \bar{1}\end{pmatrix}, \begin{pmatrix}2 & 0\\ \bar{1} & 3\end{pmatrix}$	6)

					VD	(511)	LEED-AES	$\begin{pmatrix}1 & 1\\1 & 1\end{pmatrix}, \begin{pmatrix}2 & 0\\1 & 3\end{pmatrix}$	6)
					VD	(311)	LEED-AES	$\begin{pmatrix}5 & 0\\0 & 3\end{pmatrix}$	89)
					VD	(320)	LEED-AES	$\begin{pmatrix}3 & 0\\0 & 3\end{pmatrix}$	89)
					VD	(210)	LEED-AES	$\begin{pmatrix}1 & 0\\0 & 1\end{pmatrix}$	90)
Au	Bi	rhomb	3.07	43	VD	(100)	LEED	$\begin{pmatrix}2 & 0\\\bar{1} & 2\end{pmatrix}$	91)
					VD	(111)	LEED	$\begin{pmatrix}10 & 10\\\overline{10} & 20\end{pmatrix}$	91)
					VD	(110)	LEED	$\begin{pmatrix}1 & 1\\1 & \bar{1}\end{pmatrix}, \begin{pmatrix}2 & 1\\\bar{1} & 1\end{pmatrix}, \begin{pmatrix}2 & 0\\0 & 1\end{pmatrix}$	91)
Al	–	fcc	2.86	68	–	–	–	–	–
Al	Na	bcc	3.66	24	ID	(100)	LEED-AES-WF	$\begin{pmatrix}1 & 1\\1 & \bar{1}\end{pmatrix}, \begin{pmatrix}2 & 0\\\bar{1} & 4\end{pmatrix}$	92-94)
					ID	(111)	LEED-AES-WF	$\begin{pmatrix}1 & 1\\\bar{1} & 2\end{pmatrix}, \begin{pmatrix}2 & 0\\0 & 1\end{pmatrix}$	93)
Al	Mn	cubic	2.24	54	VD	(111)	LEED-AES	$\begin{pmatrix}6 & 0\\\bar{1} & 2\end{pmatrix}$, hexagonal rotated ±9°	95)
Al	Fe	bcc	2.48	85	VD	(100)	TED	poor epitaxy	13)
Al	Ni	fcc	2.49	91	VD	(111)	TED	$\begin{pmatrix}1 & 1\\\bar{1} & 2\end{pmatrix}$	96)
Al	Sn	diam	2.81	70	VD	(100)	LEED-AES	$\begin{pmatrix}2 & 0\\\bar{1} & 3\end{pmatrix}$	97)
					VD	(111)	LEED-AES	hexagonal rotated ±9°	98)
Al	Pb	fcc	3.50	42	VD	(100)	LEED-AES	$\begin{pmatrix}2 & 0\\\bar{1} & 2\end{pmatrix}$	97)
					VD	(111)	LEED-AES	hexagonal rotated ±9°	98)
Nb	–	bcc	2.86	172	–	–	–	–	–
Nb	Sn	diam	2.81	70	VD	(110)	LEED	disordered structures, $\begin{pmatrix}3 & 0\\0 & 1\end{pmatrix}$	99)
Ta	–	bcc	2.86	180	–	–	–	–	–
Ta	Au	fcc	2.88	82	VD	(100)	LEED	split $\begin{pmatrix}1 & 1\\\bar{1} & 2\end{pmatrix}$	100)
Ta	Al	fcc	2.86	68	VD	(110)	LEED	hexagonal, square	101)

Table 5.1 (continued)

Substrate metal	Adsorbed metal	Structure	Nearest neighbor distance	Heat of vaporization (kcal/g atom)	Deposition technique	Substrate orientation	Technique of investigation	Surface structures observed	References
Ta	Th	fcc	3.60	137	VD	(100)	LEED-WF	$\begin{pmatrix}1 & 1\\ 1 & \bar{1}\end{pmatrix}$, $\begin{pmatrix}1 & 0\\ 0 & 1\end{pmatrix}$	102)
Mo	–	bcc	2.72	128	–	–	–	–	–
Mo	Na	bcc	3.66	24	ID	(110)	LEED-AES	no ordered structure	103)
Mo	K	bcc	4.52	19	ID	(110)	LEED-AES	hexagonal	103)
Mo	Rb	bcc	4.84	18	ID	(110)	LEED-AES/ AES	hexagonal	103, 104)
Mo	Cs	bcc	5.23	16	ID	(110)	LEED-AES	hexagonal	103)
Mo	Ag	fcc	2.89	61	VD	(100)	SEM-AES/ LEED-AES	(100)Ag ‖ (100)Mo and ⟨011⟩Ag ‖ ⟨001⟩Mo	105-106)
Mo	Al	fcc	2.86	68	VD	(110)	LEED-AES	hexagonal	107)
Mo	Sn	rhomb	2.81	70	VD	(100)	LEED-AES	$\begin{pmatrix}1 & 1\\ 1 & \bar{1}\end{pmatrix}$, $\begin{pmatrix}1 & 0\\ 0 & 2\end{pmatrix}$	108)
W	–	bcc	2.74	185	–	–	–	–	–
W	Li	bcc	3.02	32	VD	(110)	LEED-WF	$\begin{pmatrix}1 & 5\\ \bar{2} & 2\end{pmatrix}$, $\begin{pmatrix}2 & 0\\ 0 & 2\end{pmatrix}$, $\begin{pmatrix}1 & 1\\ \bar{1} & 2\end{pmatrix}$	109-111)
					VD	(112)	LEED-WF	$\begin{pmatrix}4 & 0\\ 0 & 1\end{pmatrix}$, $\begin{pmatrix}3 & 0\\ 0 & 1\end{pmatrix}$, $\begin{pmatrix}2 & 0\\ 0 & 1\end{pmatrix}$, incoherent, $\begin{pmatrix}1 & 0\\ 0 & 1\end{pmatrix}$	110, 112)
W	Na	bcc	3.66	24	VD	(100)	RHEED-TD	$\begin{pmatrix}1 & 1\\ 1 & \bar{1}\end{pmatrix}$	113)
					VD	(110)	LEED-WF	$\begin{pmatrix}1 & 5\\ \bar{2} & 2\end{pmatrix}$, $\begin{pmatrix}2 & 0\\ 0 & 2\end{pmatrix}$, $\begin{pmatrix}1 & 1\\ \bar{1} & 2\end{pmatrix}$, $\begin{pmatrix}1 & 1\\ 0 & 8\end{pmatrix}$, $\begin{pmatrix}1 & 1\\ 0 & 5\end{pmatrix}$, hexagonal	7)
					ID	(112)	LEED	$\begin{pmatrix}2 & 0\\ 0 & 1\end{pmatrix}$, compressed $\begin{pmatrix}2 & 0\\ 0 & 1\end{pmatrix}$	114)
W	K	bcc	4.52	19	VD	(100)	RHEED	$\begin{pmatrix}1 & 1\\ 1 & 1\end{pmatrix}$	115)
W	Rb	bcc	4.84	18	ID	(100)	LEED-AES	$\begin{pmatrix}1 & 1\\ 1 & \bar{1}\end{pmatrix}$, $\begin{pmatrix}2 & 0\\ 0 & 2\end{pmatrix}$, hexagonal	116)

							LEED-WF	$\begin{pmatrix} [illegible] \\ 1 & \bar{1} \end{pmatrix}, \begin{pmatrix} [illegible] \\ 0 & 2 \end{pmatrix}$, split $\begin{pmatrix} [illegible] \\ 0 & 2 \end{pmatrix}$ $\begin{pmatrix} 2 & 0 \\ 0 & 2 \end{pmatrix}, \begin{pmatrix} 1 & 1 \\ 1 & \bar{1} \end{pmatrix}$, hexagonal	
					VD	(110)	LEED/LEED AES	disordered hexagonal, hexagonal	120, 121, 117)
W	Be	hcp	2.22	74	VD	(110)	LEED	$\begin{pmatrix} 1 & 0 \\ 0 & 9 \end{pmatrix}, \begin{pmatrix} 1 & 0 \\ 0 & 1 \end{pmatrix}, \begin{pmatrix} 9 & 0 \\ \bar{1} & 1 \end{pmatrix}$	122)
W	Sr	fcc	4.30	34	VD	(110)	LEED-WF	$\begin{pmatrix} 3 & 3 \\ \bar{2} & 5 \end{pmatrix}, \begin{pmatrix} 2 & 2 \\ 0 & 6 \end{pmatrix}, \begin{pmatrix} 2 & 2 \\ 1 & 6 \end{pmatrix}, \begin{pmatrix} 1 & 1 \\ 0 & 3 \end{pmatrix}$, hexagonal	123)
W	Ba	bcc	4.35	36	VD	(100)	LEED-WF	$\begin{pmatrix} 2 & 0 \\ \bar{8} & 2 \end{pmatrix}$, split $\begin{pmatrix} 1 & 1 \\ \bar{2} & 2 \end{pmatrix}, \begin{pmatrix} 1 & 1 \\ \bar{2} & 2 \end{pmatrix}, \begin{pmatrix} 1 & 1 \\ 1 & \bar{1} \end{pmatrix}$	124)
					VD	(110)	LEED-WF	disordered hexagonal, hexagonal, $\begin{pmatrix} 2 & 2 \\ 0 & 6 \end{pmatrix}, \begin{pmatrix} 2 & 2 \\ 0 & 5 \end{pmatrix}, \begin{pmatrix} 3 & 3 \\ 1 & 5 \end{pmatrix}$, hexagonal compact	125)
W	Sc	hcp	3.25	81	VD	(110)	LEED-WF	$\begin{pmatrix} 1 & 1 \\ 0 & 3 \end{pmatrix}, \begin{pmatrix} 2 & 2 \\ 0 & 8 \end{pmatrix}$	126)
W	Y	hcp	3.55	93	VD	(110)	LEED-WF	hexagonal	127, 128)
W	Zr	hcp	3.17	122	VD	(100)	LEED-RHEED	$\begin{pmatrix} 1 & 0 \\ 0 & 1 \end{pmatrix}, \begin{pmatrix} 2 & 0 \\ \bar{1} & 2 \end{pmatrix}$	129)
W	Fe	bcc	2.48	85	VD	(110)	LEED	three dimensional crystals	10)
W	Pd	fcc	2.75	90	VD	(110)	LEED-AES	$\begin{pmatrix} 1 & 0 \\ 0 & 3 \end{pmatrix}$, hexagonal	130)
W	Cu	fcc	2.56	73	VD	(100)	LEED-AES-TD	$\begin{pmatrix} 2 & 0 \\ 0 & 2 \end{pmatrix}, \begin{pmatrix} 1 & 1 \\ 1 & \bar{1} \end{pmatrix}$	131)
					VD	(110)	LEED/LEED-AES-WF-TD	hexagonal structures	132, 131, 133)
W	Ag	fcc	2.89	61	VD	(100)	LEED-AES-WF-TD	$\begin{pmatrix} 2 & 0 \\ 0 & 1 \end{pmatrix}, \begin{pmatrix} 1 & 1 \\ 1 & \bar{1} \end{pmatrix}, \begin{pmatrix} 1 & 0 \\ 0 & 1 \end{pmatrix}$	134)
					VD	(110)	LEED-AES-WF-TD	hexagonal structures	134, 135)
W	Au	fcc	2.88	82	VD	(100)	LEED-AES-WF-TD	$\begin{pmatrix} 2 & 0 \\ 0 & 1 \end{pmatrix}, \begin{pmatrix} 2 & 0 \\ \bar{1} & 2 \end{pmatrix}, \begin{pmatrix} 1 & 0 \\ 0 & 1 \end{pmatrix}$	134)
					VD	(110)	LEED-AES-WF-TD	hexagonal structures	134, 136)

Table 5.1 (continued)

Substrate metal	Adsorbed metal	Structure	Nearest neighbor distance	Heat of vaporization (kcal/g atom)	Deposition technique	Substrate orientation	Technique of investigation	Surface structures observed	References
W	Hg	rhomb	3.01	14	VD	(100)	LEED-AES	$\begin{pmatrix}1 & 0\\0 & 1\end{pmatrix}$	137)
W	Pb	fcc	3.50	42	VD	(100)	LEED/LEED-AES-WF-TD	disordered $\begin{pmatrix}2 & 0\\0 & 2\end{pmatrix}$, split $\begin{pmatrix}1 & 1\\1 & \bar{1}\end{pmatrix}$ $\begin{pmatrix}1 & 1\\1 & \bar{1}\end{pmatrix}$, $\begin{pmatrix}1 & 1\\\bar{2} & 2\end{pmatrix}$, hexagonal/ $\begin{pmatrix}2 & 0\\0 & 2\end{pmatrix}$; $\begin{pmatrix}2 & 0\\\bar{1} & 2\end{pmatrix}$, $\begin{pmatrix}1 & 1\\1 & \bar{1}\end{pmatrix}$, $\begin{pmatrix}1 & 0\\0 & 1\end{pmatrix}$	138, 139)
					VD	(110)	LEED-AES-WF-TD	split $\begin{pmatrix}3 & 0\\\bar{1} & 1\end{pmatrix}$, $\begin{pmatrix}3 & 0\\\bar{1} & 1\end{pmatrix}$	139, 140)
W	Sb	rhomb	2.91	47	VD	(100)	RHEED	$\begin{pmatrix}2 & 0\\0 & 2\end{pmatrix}$, $\begin{pmatrix}1 & 1\\1 & \bar{1}\end{pmatrix}$, $\begin{pmatrix}1 & 0\\0 & 1\end{pmatrix}$	141, 142)
					VD	(110)	RHEED/ LEED-WF	$\begin{pmatrix}1 & 1\\0 & 4\end{pmatrix}$, $\begin{pmatrix}2 & 0\\\bar{1} & 1\end{pmatrix}$, $\begin{pmatrix}3 & 0\\\bar{1} & 1\end{pmatrix}$, $\begin{pmatrix}4 & 0\\\bar{1} & 1\end{pmatrix}$	141, 143)
					VD	(112)	RHEED	$\begin{pmatrix}2 & 0\\0 & 1\end{pmatrix}$, $\begin{pmatrix}1 & 0\\0 & 1\end{pmatrix}$	141)
W	Th	fcc	3.60	137	VD	(100)	LEED/ LEED-AES-WF	$\begin{pmatrix}1 & 1\\1 & \bar{1}\end{pmatrix}$, $\begin{pmatrix}1 & 0\\0 & 1\end{pmatrix}$/ $\begin{pmatrix}1 & 1\\1 & \bar{1}\end{pmatrix}$, $\begin{pmatrix}1 & 0\\0 & 1\end{pmatrix}$, $\begin{pmatrix}2 & 0\\\bar{1} & 3\end{pmatrix}$, hexagonal	144-146, 147, 148)
Ti	–	hcp	2.89	106	–	–	–	–	–
Ti	Cu	fcc	2.56	73	VD	(0001)	LEED	extra spots	1)
Ti	Cd	hcp	2.98	24	VD	(0001)	LEED	$\begin{pmatrix}1 & 0\\0 & 1\end{pmatrix}$	2)
Re	–	hcp	2.74	152	–	–	–	–	–
Re	Ba	bcc	4.35	36	VD	(0001)	LEED-WF	$\begin{pmatrix}2 & 0\\0 & 2\end{pmatrix}$, hexagonal	11)
Zn	–	hcp	2.66	27	–	–	–	–	–
Zn	Cu	fcc	2.56	73	VD	(0001)	LEED	$\begin{pmatrix}1 & 0\\0 & 1\end{pmatrix}$	9)
Sb	–	rhomb	2.91	62	–	–	–	–	–
Sb	Fe	bcc	2.48	85	VD	(0001)	TED	$\begin{pmatrix}1 & 0\\0 & 1\end{pmatrix}$	12)

References to Table 5.1

1. Schlier, R. E., Farnsworth, H. E.: J. Phys. Chem. Solids *6* (1958) 271
2. Shih, H. D., Jona, F., Jepsen, D. W., Marcus, P. M.: Phys. Rev. *B15* (1977) 5550;
   Shih, H. D., Jona, F., Jepsen, D. W., Marcus, P. M.: Phys. Rev. *B15* (1977) 5561;
   Shih, H. D., Jona, F., Jepsen, D. W., Marcus, P. M.: Comm. on Physics *1* (1976) 25
3. Biberian, J. P., Somorjai, G. A.: to be published.
4. Matthews, J. W., Jesser, W. A.: Acta Metall. *15* (1967) 595;
   Matthews, J. W.: Phil. Mag. *13* (1966) 1207
5. Biberian, J. P., Rhead, G. E.: J. Phys. *F3* (1973) 675;
   Biberian, J. P., Huber, M.: Surface Sci. *55* (1976) 259;
   Green, A. K., Prigge, S., Bauer, E.: Thin Solid Films, *52* (1978) 163
6. Biberian, J. P.: Surface Sci. *74* (1978) 437
7. Nedvedev, V. K., Nauvomets, A. G., Fedorus, A. G.: Sov. Phys. Solid State *12* (1970) 301
   Naumovets, A. G., Fedorus, A. G.: JETP Lett. *10* (1969) 6
8. Feinstein, L. G., Blanc, E.: Surface Sci. *18* (1969) 350
   Edmonds, T., McCarroll, J. J.: Surface Sci. *24* (1971) 353
9. Abbati, I., Braicovich, L., Bertoni, C. M., Calandra, C., Manghi, F.: Phys. Rev. Lett. *40* (1978) 469.
   Abbati, J., Braicovich, L.: Proc. 7th Vac. Congr. and 3rd Intern. Conf. Solid Surfaces (Vienna 1977), 1117.
10. Melmed, A. J., McCarroll, J. J.: Surface Sci. *19* (1970) 243
11. Gorodetskii, D. A., Knysh, A. N.: Surface Sci. *40* (1973) 636
    Gorodetskii, D. A., Knysh, A. N.: Surface Sci. *40* (1973) 651
12. Shigematsu, T., Hine, S., Takada, T.: J. Crystal Growth *43* (1978) 531.
13. Hothersall, D. C., Phil. Mag. *15* (1967) 1023
14. Thomas, R. E., Haas, G. A.: J. Appl. Phys. *43* (1972) 4900
15. Andersson, S., Kasemo, B.: Surface Sci. *32* (1972) 78
16. Gerlach, R. L., Rhodin, T. N.: Surface Sci. *17* (1969) 32
17. Andersson, S., Pendry, J. B.: J. Phys. C, *6* (1973) 601
18. Andersson, S., Jostell, U.: Surface Sci. *46* (1974) 625
19. Gerlach, R. L., Rhodin, T. N.; The structure and chemistry of solid surfaces, G. A. Somorjai, (1968), p. 55
20. Andersson, S., Pendry, J. B.: J. Phys. C, *5* (1972) L 41
21. Gerlach, R. L., Rhodin, T. N.: Surface Sci. *10* (1968) 446
22. Andersson, S., Jostell, U.: Solid State Comm. *13* (1973) 829
    Andersson, S., Jostell, U.: Solid State Comm. *13* (1973) 833
23. Papageorgopoulos, C. A., Chen, J. M.: Surface Sci. *52* (1975) 40
    Papageorgopoulos, C. A., Chen, J. M.: Surface Sci. *52* (1975) 53
24. Jesser, W. A., Matthews, J. W.: Phil. Mag. *17* (1968) 475
25. Jesser, W. A., Matthews, J. W.: Acta Mett. *16* (1968) 1307
26. Chambers, A., Jackson, D. C.: Phil. Mag. *31* (1975) 1357
27. Feinstein, L. G., Blanc, E., Dufayard, D.: Surface Sci. *19* (1970) 269
28. Jackson, D. C., Gallon, T. E., Chambers, A.: Surface Sci. *36* (1973) 381
29. Burton, J. J., Helms, C. R., Polizzotti, R. S.: Surface Sci. *57* (1976) 425
    Burton, J. J., Helms, C. R., Polizzotti, R. S.: J. Chem. Phys. *65* (1976) 1089
    Burton, J. J., Helms, C. R., Polizzotti, R. S.: J. Vac. Sci. Technol. *13* (1976) 204
30. Wolfe, J. R., Weart, H. W.: The Structure and Chemistry of Solid Surfaces, G. A. Somorjai (1968) p. 32
31. Perdereau, J., Szymerska, I.: Surface Sci. *32* (1972) 247
32. Alkhoury Nemen, E., Cinti, R. C., Nguyen, T. T. A.: Surface Sci. *30* (1972) 697
33. Matthews, J. W.: Thin Solid Films, *12* (1972) 243
34. Palmberg, P. W., Rhodin, T. N.: J. Chem. Phys. *49* (1968) 134
35. Matthews, J. W.: Phil. Mag. *13* (1966) 1207
36. Yagi, K., Takanayagi, K., Kobayashi, K., Honjo, G.: J. Crystal. Growth *9* (1971) 84
37. Jesser, W. A., Matthews, J. W.: Phil. Mag. *17* (1968) 595
38. Jesser, W. A., Matthews, J. W.: Phil. Mag. *15* (1967) 1097

References to Table 5.1 (continued)

39. Gradmann, U., Kümmerle, W., Tillmanns, P.: Thin Solid Films *34* (1976) 249
40. Jesser, W. A., Matthews, J. W.: Phil. Mag. *17* (1968) 461
41. Fedorenko, A. I., Vincent, R.: Phil. Mag. *24* (1971) 55
42. Kuntze, R., Chambers, A., Prutton, M.: Thin Solid Films *4* (1969) 47
43. Haque, C. A., Farnsworth, H. E.: Surface Sci. *4* (1966) 195
44. Gradmann, U.: Surface Sci. *13* (1969) 498
45. Gradmann, U.: Ann. Physik. *13* (1964) 213. Gradmann, U.: Ann. Physik. *17* (1966) 91
46. Bauer, E.: Surface Sci. *7* (1967) 351
47. Vook, R. W., Horng, C. T., Macur, J. E.: J. of Crystal Growth *31* (1975) 353
48. Vook, R. W., Horng, C. T.: Phil. Mag. *33* (1976) 843
49. Horng, C. T., Vook, R. W.: Surface Sci. *54* (1976) 309
50. Gradmann, U.: Phys. Kondens. Materie *3* (1964) 91
51. Palmberg, P. W., Rhodin, T. N · J. of Appl. Phys. *39* (1968) 2425
52. Fujinaga, Y.: Surface Sci. *64* (1977) 751
53. Vook, R. W., Macur, J. E.: Thin Solid Films *32* (1976) 199.
54. Erlewein, J., Hofmann, S.: Surface Sci. *68* (1977) 71
55. Henrion, J., Rhead, G. E.: Surface Sci. *29* (1972) 20
56. Sepulveda, A., Rhead, G. E.: Surface Sci. *66* (1977) 436
57. Argile, C., Rhead, G. E.: Surface Sci. *78* (1978) 115
58. Barthes, M. G., Rhead, G. E.: Surface Sci. *80* (1979) 421.
59. Reichelt, K., Müller, F.: J. of Crystal Growth *21* (1974) 323
60. Delamare, F., Rhead, G. E.: Surface Sci. *35* (1973) 172
61. Delamare, F., Rhead, G. E.: Surface Sci. *35* (1973) 185
62. Goddard, P. J., West, J., Lambert, R. M.: Surface Sci. *71* (1978) 447
63. Marbrow, R. A., Lambert, R. M.: Surface Sci. *61* (1976) 329
64. Goddard, P. J., Lambert, R. M.: Surface Sci. *79* (1979) 93
65. Newman, R. C.: Phil. Mag. *2* (1957) 750
66. Bruce, L. A., Jaeger, H.: Phil. Mag. *36* (1977) 1331
67. Gonzalez, C.: Acta Mett. *15* (1967) 1373
68. Horng, C. T., Vook, R. W.: J. Vac. Sci. Technol. *11* (1974) 140
69. Grünbaum, E., Kremer, G., Reymond, C.: J. Vac. Sci. Technol. *6* (1969) 475
70. Newman, R. C., Pashley, D. W.: Phil. Mag. *46* (1955) 927
71. Soria, F., Sacedon, J. L., Echenique, P. M., Titterington, D.: Surface Sci. *68* (1977) 448.
72. Klaua, M., Bethge, H.: J. of Crystal Growth *3, 4* (1968) 188
73. Grunbaum, E.: Proc. Phys. Soc. (London) *72* (1958) 459
74. Bauer, E.: Structure et Propriétés des Solides, CNRS, Paris (1969)
75. Thomas, R. E., Haas, G. A.: J. Appl. Phys. *43* (1972) 4900
76. Wassermann, E. F. Jablonski, H. P.: Surface Sci. *22* (1970) 69
77. Hothersall, D. C.: Phil. Mag. *15* (1967) 1023
78. Gueguen, P., Camoin, C., Gillet, M.: Thin Solid Films *26* (1975) 107
79. Honjo, G., Takayanagi, K., Kobayashi, K., Yagi, K.: J. of Crystal Growth *42* (1977) 98
80. Gueguen, P., Cahareau, M., Gillet, M.: Thin Solid Films *16* (1973) 27
81. Cherns, D., Stowell, M. J.: Thin Solid Films *29* (1975) 107
82. Cherns, D., Stowell, M. J.: Thin Solid Films *29* (1975) 127
83. Jesser, W. A., Matthews, J. W., Kuhlmann-Wilsdorf, D.: Appl. Phys. Lett. *9* (1966) 176
84. Macur, J. E., Vook, R. W.: 32nd Ann. Proc. Electron Microscopy Soc. Amer. St Louis, Missouri, 1974, C. J. Arceneaux (ed.).
85. Macur, J. E.: 33rd Ann. Proc. Electron Microscopy Soc. Amer., Las Vegas, Nevada, 1975, B. W. Bailey (ed.), 98
86. Farnsworth, H. E.: Phys. Rev. *40* (1932) 684
87. Matthews, J. W.: Phys. Thin Films *4* (1967) 137
88. Perdereau, J., Biberian, J. P., Rhead, G. E.: J. Phys. F *4* (1974) 798
89. Barthes, M. G., Rhead, G. E.: Surface Sci., to be published
90. Barthes, M. G.: Thesis University of Paris, 1978
91. Sepulveda, A., Rhead, G. E.: Surface Sci. *49* (1975) 669
92. Hutchins, B. A., Rhodin, T. N., Demuth, J. E.: Surface Sci. *54* (1976) 419
93. Porteus, J. O.: Surface Sci. *41* (1974) 515
94. Van Hove, M., Tong, S. Y., Stoner, N.: Surface Sci. *54* (1976) 259
95. Edwards, I. A. S., Thirsk, H. R.: Surface Sci. *39* (1973) 245

References to Table 5.1 (continued)

96. Dorey, G.: Thin Solid Films *5* (1970) 69
97. Argile, C., Rhead, G. E.: Surface Sci. *78* (1978) 125
98. Argile, C.: Thesis University of Paris, 1978
99. Jackson A. G., Hooker, M. P.: "The structure and chemistry of solid surfaces", G. A. Somorjai (1968) p. 73
100. Elliot, A. G.: Surface Sci. *51* (1975) 489
Biberian, J. P.: Surface Sci. *59* (1976) 307
101. Haaš, T. W., Jackson, A. G., Hooker, M. P.: J. Appl. Phys. *38* (1967) 4998
Jackson, A. G., Hooker, M. P., Haas, T. W.: Surface Sci. *10* (1968) 308
102. Pollard, J. H., Danforth, W. E.: "The structure and chemistry of solid surfaces", G. A. Somorjai (1968) p. 71
Pollard, J. H., Danforth, W. E.: J. Appl. Phys. *39* (1968) 4019
103. Thomas, S., Haas, T. W.: J. Vac. Sci. Techn. *9* (1972) 840
104. Thomas, S., Haas, T. W.: Surface Sci. *28* (1971) 632
105. Hartig, K., Janssen, A. P., Venables, J. A.: Surface Sci. *74* (1978) 69
106. Hartig, K.: Thesis Ruhr-Universität, Bochum
107. Jackson, A. G., Hooker, M. P.: Surface Sci. *28* (1971) 373
108. Jackson, A. G., Hooker, M. P., Surface Sci. *27* (1971) 197
109. Gorodetsky, D. A., Melnik, Yu. P., Yasko, A. A.: Ukr. Fiz, Zhurn. *12* (1967) 649
110. Medvedev, V. K., Smereka, T. P.: Sov. Phys. Solid State *16* (1974) 1046
111. Naumovets, A. G., Fedorus, A. G.: Sov. Phys. JETP *41* (1975) 587
112. Medvedev, V. K., Naumovets, A. G., Smereka, T. P.: Surface Sci. *34* (1973) 368
113. Mlynczak, A., Niedermayer, R.: Thin Solid Films *28* (1975) 37
114. Chen, J. M., Papageorgopoulos, C. A.: Surface Sci. *21* (1970) 377
115. Steinhage, P. W., Mayer, H.: Thin Solid Films, *28* (1975) 131
116. Thomas, S., Haas, T. W.: J. Vac. Sci. Technol. *10* (1973) 218
117. Mac Rae, A. U., Müller, K., Lander, J. J., Morrison, J.: Surface Sci. *15* (1969) 483
118. Papageorgopoulos, C. A., Chen, J. M.: Surface Sci. *39* (1973) 283
119. Voronin, V. B., Nauvomets, A. G., Fedorus, A. G.: JETP lett. *15* (1972) 370
Wang, C. S.: J. Appl. Phys. *48* (1977) 1477
120. Fedorus, A. G., Naumovets, A. G.: Surface Sci. *21* (1970) 426
121. Fedorus, A. G., Naumovets, A. G.: Sov. Phys. Solid State *12* (1970)
122. Niehus, H.: Thesis Clausthal, 1975
123. Kanash, O. V., Naumovets, A. G., Fedorus, A. G.: Sov. Phys. JETP *40* (1974) 903
124. Gorodetskii, D. A., Mel'nik, Yu. P.: Akad. Nauk SSSR *33* (1969) 430
Gorodetskii, D. A., Mel'nik, Yu. P., Sklyar, V. A., Usenko, V. A.: Surface Sci., to be published.
125. Gorodetskii, D. A., Mel'nik, Yu. P.: Surface Sci. *62* (1977) 647
Gorodetskii, D. A., Gorchinskii, A. D., Maksimenko, V. I., Mel'nik, Yu. P.: Sov. Phys. Solid State *18* (1976) 691
Gorodetskii, D. A., Kornev, A. M., Mel'nik, Yu. P.: Izv. Akad. Nauk, SSSR, Ser. Fiz. *28* (1964) 1337
126. Voronin, V. B., Naumovets, A. G.: Ukr. Fiz. Zhurn. *13* (1968) 1389
Voronin, V. B.: Soviet Phys. Solid State *9* (1968) 1758
Gorodetskii, D. A., Yas'ko, A. A.: Sov. Phys. Solid State *10* (1969) 1812
127. Gorodetskii, D. A., Yas'ko, A. A., Shevlyakov, S. A.: Izv. Akad. Nauk SSSR, Ser. Fiz. *35* (1971) 436
128. Voronin, V. B., Naumovets, A. G.: Izv. Akad. Nauk SSSR Ser. Fiz. *35* (1971) 325
129. Hill, G. E., Marklund, J., Martinson, J.: Surface Sci. *24* (1971) 435
130. Paraschkevov, D., Schlenk, W., Bajpai, R. P., Bauer, E.: Proc. 7th Intern. Vac. Congr. and 3rd Intern. Conf. Solid Surfaces, Vienna (1977) 1737
131. Bauer, E., Poppa, H., Todd, G., Bonczek, F.: J. Appl. Physics *45* (1974) 5164
132. Taylor, N. J.: Surface Sci. *4* (1966) 161

References to Table 5.1 (continued)

133. Moss. A. R., Blott, B. H.: Surface Sci. *17* (1969) 240
134. Bauer, E., Poppa, H., Todd, G., Davis, P. R.: J. Appl. Physics *48* (1977), 3773
135. Hudson, J. B., Lo, C. M.: Surface Sci. *36* (1973) 141
136. Augustus, P. D., Jones, J. P.: Surface Sci. *64* (1977) 713
137. Jones, R. G., Perry, D. L.: Surface Sci. *71* (1978) 59
138. Gorodetskii, D. A., Yas'ko, A. A.: Sov. Phys. Solid State *14* (1972) 636
139. Bauer, E., Poppa, H., Todd, G.: Thin Solid Films *28* (1975) 19
140. Gorodetskii, D. A., Yas'ko, A. A.: Sov. Phys. Solid State *11* (1969) 640
141. Hopkins, B. J., Watts, G. D.: Surface Sci. *47* (1975) 195
142. Hopkins, B. J., Watts, G. D.: Surface Sci. *45* (1974) 77
143. Gorodetskii, D. A., Yas'ko, A. A.: Sov. Phys. Solid State *13* (1971) 1085
144. Estrup, P. J., Anderson, J., Danforth, W. E.: Surface Sci. *4* (1966) 286
145. Estrup, P. J., Anderson, J.: Surface Sci. *7* (1967) 255
146. Estrup, P. J., Anderson, J.: Surface Sci. *8* (1967) 101
147. Pollard, J. H.: Surface Sci. *20* (1970) 269
148. Anderson, J., Estrup, P. J., Danforth, W. E.: Appl. Phys. Lett. *7* (1965) 122

Table 5.2. Surface structures on substrates with threefold rotational symmetry

Surface	Adsorbed gas	Surface structure	References
Ag(111)	O_2	$(2 \times 2) - O$	1)
		$(\sqrt{3} \times \sqrt{3})R30° - O$	1)
		Not adsorbed	146)
		$(4 \times 4) - O$	147, 148)
	I_2	$(\sqrt{3} \times \sqrt{3})R30° - I$	145, 149, 150)
	Cl_2	$(\sqrt{3} \times \sqrt{3})R30° - Cl$	151)
		$(10 \times 10) - Cl$	151)
		AgCl(111)	152)
	$C_2H_4Cl_2$	$(\sqrt{3} \times \sqrt{3})R30° - Cl$	153, 154)
		$(3 \times 3) - Cl$	153, 154)
	Br_2	$(\sqrt{3} \times \sqrt{3})R30° - Br$	155)
		$(3 \times 3) - Br$	155)
	Xe	Hexagonal overlayer	156, 157, 158, 159, 160)
	Kr	Hexagonal overlayer	156)
	$CO + O_2$	$(2 \times \sqrt{3}) - (CO + O_2)$	27)
	NO	Disordered	163)
Al(111)	O_2	$(4 \times 4) - O$	123)
Au(111)	O_2	Oxide	161)
		Not adsorbed	162)
		Adsorbed	162)
Be(0001)	O_2	Disordered	22)
	CO	Disordered	22)
	H_2	Not adsorbed	22)
	N_2	Not adsorbed	22)
C(111), diamond	O_2	Adsorbed	16)
		Not adsorbed	164)
	N_2	Not adsorbed	164)
	NH_3	Not adsorbed	164)
	H_2S	Not adsorbed	164)
	H_2	$(1 \times 1) - H$	30)
	P	$(\sqrt{3} \times \sqrt{3})R30° - P$	30)
C(0001), graphite	Xe	$(\sqrt{3} \times \sqrt{3})R30° - Xe$	165)
	Kr	$(\sqrt{3} \times \sqrt{3})R30° - Kr$	166, 167, 174)
CdS(0001)	O_2	Disordered	25)
Co(0001)	CO	$(\sqrt{3} \times \sqrt{3})R30° - CO$	168)
		Hexagonal overlayer	168)
Cr(111)	O_2	$(\sqrt{3} \times \sqrt{3})R30° - O$	169)
Cu(111)	O_2	Disordered	7, 170, 171)
		$(7 \times 7) - O$	7, 8)
		$(\sqrt{3} \times \sqrt{3})R30° - O$	7, 8)
		$(2 \times 2) - O$	115, 7, 8)
		$(3 \times 3) - O$	8)
		$(11 \times 5)R5° - O$	9)
		$(2 \times 2)R30° - O$	115, 119)
		Hexagonal	246)
	CO	Not adsorbed	26)
		$(\sqrt{3} \times \sqrt{3})R30°$	172, 173)
		$(\sqrt{7/3} \times \sqrt{7/3})R49.1°$	172, 173)
		$(3/2 \times 3/2)$	173)

Table 5.2 (continued)

Surface	Adsorbed gas	Surface structure	References
	Cl_2	$(\sqrt{3} \times \sqrt{3})R30° - Cl$	151)
		$(6\sqrt{3} \times 6\sqrt{3})R30° - Cl$	151)
		$(12\sqrt{3} \times 12\sqrt{3})R30° - Cl$	151)
		$(4\sqrt{7} \times 4\sqrt{7})R19.2° - Cl$	151)
	H_2	Not adsorbed	7)
	H_2S	$(\sqrt{3} \times \sqrt{3})R30° - S$	35)
		Adsorbed	35)
	Xe	$(\sqrt{3} \times \sqrt{3})R30° - Xe$	159)
Cu/Ni(111)	CO	Disordered	173)
Fe(111)	O_2	$(6 \times 6) - O$	175)
		$(5 \times 5) - O$	175)
		$(4 \times 4) - O$	175)
		$(2\sqrt{7} \times 2\sqrt{7})R19.1° - O$	175)
		$(2\sqrt{3} \times 2\sqrt{3})R30° - O$	175)
	NH_3	Disordered	176)
		$(3 \times 3) - N$	176)
		$(\sqrt{19} \times \sqrt{19})R23.4° - N$	176)
		$(\sqrt{21} \times \sqrt{21})R10.9° - N$	176)
	H_2	Adsorbed	177)
Ge(111)	O_2	Disordered	17, 18)
		(1×1)	19, 21)
	P	$(1 \times 1) - P$	19)
	H_2S	$(2 \times 2) - S$	37)
		$(2 \times 1) - S$	178)
	H_2Se	$(2 \times 2) - Se$	37)
	H_2O	$(1 \times 1) - H_2O$	121, 179)
	I_2	$(1 \times 1) - I$	19)
Ir(111)	O_2	$(2 \times 2) - O$	124, 180, 181, 182, 183, 184)
		$(2 \times 1) - O$	182)
		Ir oxide	181)
	CO	$(\sqrt{3} \times \sqrt{3})R30° - CO$	124, 180, 182, 183, 185, 186)
		$(2\sqrt{3} \times 2\sqrt{3})R30° - CO$	180, 182, 183, 185, 186)
	H_2O	Not adsorbed	182)
	H_2	Adsorbed	187)
	NO	$(2 \times 2) - NO$	188)
Mo(111)	O_2	(211) facets	14, 189)
		(110) facets	189)
		$(4 \times 2) - O$	190)
	H_2S	$C(4 \times 2) - H_2S$	191)
		$MoS_2(0001)$	191)
Nb(111)	O_2	$(2 \times 2) - O$	192)
		$(1 \times 1) - O$	192)
Ni(111)	O_2	$(2 \times 2) - O$	2, 3, 4, 116, 193, 194, 195, 196, 197, 198)
		$(\sqrt{3} \times \sqrt{3})R30° - O$	2, 5, 195)
		$(\sqrt{3} \times \sqrt{21}) - O$	116)
		NiO(111)	4, 6, 116, 193, 194)

Table 5.2 (continued)

Surface	Adsorbed gas	Surface structure	References
	CO	$(\sqrt{3} \times \sqrt{3})R30° - CO$	195, 196, 199, 200)
		Hexagonal overlayer	200)
		$(2 \times 2) - CO$	3)
		$(\sqrt{3} \times \sqrt{3})R30° - O$	5)
		$(2 \times \sqrt{3}) - CO$	5)
		$(\sqrt{39} \times \sqrt{39}) - C$	5, 27)
		Disordered	198)
		$(\sqrt{7} \times \sqrt{7})R19.1°$	195, 196)
		$C(4 \times 2)$	195, 196)
	CO_2	$(2 \times 2) - CO_2$	5)
		$(\sqrt{3} \times \sqrt{3})R30° - O$	5)
		$(2 \times \sqrt{3}) - CO_2$	5)
		$(\sqrt{39} \times \sqrt{39}) - C$	5, 27)
	H_2	$(1 \times 1) - H$	3)
		$(2 \times 2) - H$	29, 201, 202, 204)
		Disordered	203)
	NO	$C(4 \times 2) - NO$	193)
		Hexagonal overlayer	193)
		$(2 \times 2) - O$	193)
		$(6 \times 2) - N$	193)
	H_2S	$(2 \times 2) - S$	36, 118, 197, 198, 205, 294)
		$(\sqrt{3} \times \sqrt{3})R30° - S$	36, 118)
		$(5 \times 5) - S$	36)
		adsorbed	36)
	H_2Se	$(2 \times 2) - Se$	137)
		$(\sqrt{3} \times \sqrt{3})R30° - Se$	137)
	Cl_2	$(\sqrt{3} \times \sqrt{3})R30° - Cl$	206)
		$\begin{pmatrix} 2 & 1 \\ 4 & 7 \end{pmatrix} - Cl$	206)
	N_2	Not adsorbed	131)
Pd(111)	O_2	$(2 \times 2) - O$	207)
		$(\sqrt{3} \times \sqrt{3})R30° - O$	207)
		$(2 \times 2) - PdO$	207)
	NO	$C(4 \times 2) - NO$	208)
		$(2 \times 2) - NO$	208)
	CO	$(\sqrt{3} \times \sqrt{3})R30° - CO$	209, 210)
		Hexagonal overlayer	209)
		$C(4 \times 2) - CO$	210)
	H_2	$(1 \times 1) - H$	211, 212)
Pt(111)	O_2	$(2 \times 2) - O$	10, 11, 213, 214, 215, 216, 217)
		$(\sqrt{3} \times \sqrt{3}) - R30° - O$	214, 215, 217)
		Not adsorbed	120)
		$(4\sqrt{3} \times 4\sqrt{3})R30° - O$	214, 215)
		$PtO_2(0001)$	214, 215)
		$(3 \times 15) - O$	217)
	CO	$(\sqrt{3} \times \sqrt{3})R30° - CO$	218)
		$C(4 \times 2) - CO$	28, 120, 218, 219)

Table 5.2 (continued)

Surface	Adsorbed gas	Surface structure	References
		Hexagonal overlayer	218)
		$(2 \times 2) - CO$	120)
	H_2	Not adsorbed	120)
		Adsorbed	220, 221)
	$H_2 + O_2$	$(\sqrt{3} \times \sqrt{3})R30°$	11)
	NO	Disordered	222)
	H_2O	$(\sqrt{3} \times \sqrt{3})R30° - H_2O$	223, 224)
		$H_2O(111)$	224)
	S_2	$(2 \times 2) - S$	225, 226, 227, 247)
		$(\sqrt{3} \times \sqrt{3})R30° - S$	225, 226, 227, 247)
		$\begin{pmatrix} 4 & -1 \\ -1 & 2 \end{pmatrix} - S$	225, 226)
		Hexagonal	227)
	N	Disordered	228)
Re(0001)	O_2	$(2 \times 2) - 0$	23, 24, 229)
	CO	Not adsorbed	24)
		$(2 \times 2) - CO$	23)
		Disordered	230)
		$(2 \times \sqrt{3})$	230)
	H_2	Not adsorbed	24)
	N_2	Not adsorbed	24)
Rh(111)	O_2	$(2 \times 2) - O$	12, 231)
	CO	$(\sqrt{3} \times \sqrt{3})R30° - CO$	231)
		$(2 \times 2) - CO$	12, 231)
	CO_2	$(\sqrt{3} \times \sqrt{3})R30° - CO$	231)
		$(2 \times 2) - CO$	231)
	H_2	Adsorbed	231)
	NO	$C(4 \times 2) - NO$	231)
		$(2 \times 2) - NO$	231)
Ru(0001)	O_2	$(2 \times 2) - O$	12, 232, 248)
	CO	$(\sqrt{3} \times \sqrt{3})R30° - CO$	12, 233, 248)
		$(2 \times 2) - CO$	12, 248)
	CO_2	$(\sqrt{3} \times \sqrt{3})R30° - CO_2$	12)
		$(2 \times 2) - CO_2$	12)
	H_2	$(1 \times 1) - H$	234)
	N_2	Adsorbed	234)
	NH_3	$(2 \times 2) - NH_3$	234, 235)
		$(\sqrt{3} \times \sqrt{3})R30° - NH_3$	235)
Si(111)	O_2	Disordered	17, 20, 21)
	N_2	$(8 \times 8) - N$	34)
	P	$(6\sqrt{3} \times 6\sqrt{3}) - P$	132, 133)
		$(1 \times 1) - P$	132)
		$(2\sqrt{3} \times 2\sqrt{3}) - P$	132)
		$(4 \times 4) - P$	133)
	Cl_2	Disordered	138)
		$(7 \times 7) - Cl$	138, 236)
		$(1 \times 1) - Cl$	138, 236)
	I_2	$(1 \times 1) - I$	133)

Table 5.2 (continued)

Surface	Adsorbed gas	Surface structure	References
	H_2	$(1 \times 1) - H$	237)
		$(7 \times 7) - H$	237)
	NH_3	$(8 \times 8) - N$	238)
	PH_3	$(7 \times 7) - P$	239)
		$(1 \times 1) - P$	239)
		$(6\sqrt{3} \times 6\sqrt{3}) - P$	239)
		$(2\sqrt{3} \times 2\sqrt{3}) - P$	239)
Ti(0001)	O_2	$(1 \times 1) - O$	18)
	CO	$(1 \times 1) - CO$	18, 240)
		$(2 \times 2) - CO$	240)
	N_2	$(1 \times 1) - N$	241, 242)
		$(\sqrt{3} \times \sqrt{3})R30° - N$	241, 242)
Th(111)	O_2	Disordered	243)
		$ThO_2(111)$	243)
	CO	Disordered	243)
		$ThO_2(111)$	243)
$UO_2(111)$	O_2	$(3 \times 3) - O$	13)
		$(2\sqrt{3} \times 2\sqrt{3})R30° - O$	13)
W(111)	O_2	Disordered	244)
		(211) facets	15)
Zn(0001)	O_2	$(1 \times 1) - O$	122)
		ZnO(0001)	245)
Zn(000$\bar{1}$)	O_2	$(\sqrt{3} \times \sqrt{3})R30° - O$	122)

Table 5.3. Surface structures on substrates with fourfold rotational symmetry

Surface	Adsorbed gas	Surface structure	References
Ag(100)	O_2	Disordered	146)
	$C_2H_4Cl_2$	$C(2 \times 2) - Cl$	154, 249)
	Se	$C(2 \times 2) - Se$	250)
Al(100)	O_2	Disordered	42, 43, 44)
Au(100)	H_2S	$(2 \times 2) - S$	251)
		$C(2 \times 2) - S$	251)
		$(6 \times 6) - S$	251)
		$C(4 \times 4) - S$	251)
	CO	Disordered	252)
	Xe	Disordered	252)
C(100), diamond	O_2	Disordered	16)
		Not adsorbed	164)
	N_2	Not adsorbed	164)
	NH_3	Not adsorbed	164)
	H_2S	Not adsorbed	164)
Co(100)	CO	$C(2 \times 2) - CO$	253)
		$(2 \times 2) - C$	253)
	O_2	$(2 \times 2) - O$	254)
		$C(2 \times 2) - O$	254)

Table 5.3 (continued)

Surface	Adsorbed gas	Surface structure	References
Cr(100)	O_2	C(2 × 2) – O	255)
		Cr_2O_3(310)	256)
Cu(100)	O_2	(1 × 1) – O	9, 45)
		(2 × 1) – O	9, 45, 46)
		(2 × 4)R45° – O	7, 47, 246, 261)
		(2 × 3) – O	119)
		C(4 × 4) – O	119)
		C(2 × 2) – O	171, 246, 257, 258, 259, 260, 263, 264)
		(2 × 2)	171)
		(2 × $2\sqrt{2}$)R45°	259, 262, 263, 264)
		Hexagonal	259)
		(410) facets	259)
	CO	C(2 × 2) – CO	125, 126, 265)
		Hexagonal overlayer	126, 127, 265)
		(2 × 2) – C	26, 125)
	N_2	(1 × 1) – N	49)
		C(2 × 2) – N	47, 132, 258, 261, 266)
Mo(100)	O_2	Disordered	61, 62)
		C(2 × 2) – O	61, 62, 63, 64, 284)
		($\sqrt{5}$ × $\sqrt{5}$)R26° – O	61, 62, 189, 190, 284)
		(2 × 2) – O	61, 189, 190)
		C(4 × 4) – O	62, 189, 284)
		(2 × 1) – O	189, 190)
		(6 × 2) – O	284)
		(3 × 1) – O	284)
		(1 × 1) – O	284)
	CO	Disordered	62)
		(1 × 1) – CO	62, 64, 285, 286)
		C(2 × 2) – CO	64, 285, 286)
		(4 × 1) – CO	64)
	H_2	C(4 × 2) – H	77)
		(1 × 1) – H	77)
	N_2	(1 × 1) – N	62)
		C(2 × 2) – N	287)
	H_2S	(1 × 1) – S	130)
		($\sqrt{5}$ × $\sqrt{5}$) – S	130, 288)
		C(2 × 2) – S	130)
		MoS_2(100)	288)
NaCl(100)	Xe	Hexagonal overlayer	289)
Nb(100)	O_2	C(2 × 2) – O	192, 290)
		(1 × 1) – O	192, 290)
		(3 × 10) – NbO_2	290)
	N_2	(5 × 1) – N	290)
Ni(100)	O_2	(2 × 2) – O	2, 49, 50, 51, 198, 296-299, 310
		C(2 × 2) – O	2, 6, 52-57, 197, 198, 290-299, 310, 340)

Table 5.3 (continued)

Surface	Adsorbed gas	Surface structure	References
		$(2 \times 1) - O$	198)
		NiO(100)	6, 297, 298, 299, 310)
		NiO(111)	298, 299)
	CO	$C(2 \times 2) - CO$	54, 55, 68, 129, 198, 300, 301, 302)
		$(2 \times 2) - CO$	69)
		Hexagonal overlayer	129, 301, 302)
		$(2 \times 2) - C$	198)
	CO_2	$(2 \times 2) - O + C(2 \times 2) - CO$	76)
	N_2	Not adsorbed	80, 81)
	H_2	Disordered	198, 203, 211)
		$C(2 \times 2) - H$	301)
	H_2S	Adsorbed	35)
		$(2 \times 2) - S$	35, 260, 262)
		$(2 \times 1) - S$	128)
	Te	$(2 \times 2) - Te$	267)
	Xe	Hexagonal overlayer	159)
Fe(100)	O_2	$C(2 \times 2) - O$	60, 269, 270, 271, 274)
		$(1 \times 1) - O$	144, 268, 271, 272)
		FeO(100)	60, 269, 270, 272, 273)
		FeO(111)	270)
		FeO(110)	272)
		Disordered	273, 276)
	CO	$C(2 \times 2) - CO$	275)
	H_2S	$C(2 \times 2) - S$	276, 277)
	H_2	Adsorbs	177)
	NH_3	Disordered	176)
		$C(2 \times 2) - N$	176)
	H_2O	$C(2 \times 2)$	278)
Fe/Cr(100)	O_2	$C(2 \times 2) - O$	279)
		$C(4 \times 4) - O$	279)
		Oxide	280)
Ge(100)	O_2	Disordered	17, 18)
	I_2	$(3 \times 3) - I$	19)
Ir(100)	O_2	$(2 \times 1) - O$	48, 281)
		$(5 \times 1) - O$	48)
	CO	$C(2 \times 2) - CO$	48)
		$(2 \times 2) - CO$	48)
		$(1 \times 1) - CO$	282)
	CO_2	$C(2 \times 2) - CO_2$	48)
		$(2 \times 2) - CO_2$	48)
		$(7 \times 20) - CO_2$	48)
	NO	$(1 \times 1) - NO$	188)
	H_2	Adsorbed	281)
	Kr	$(3 \times 5) - Kr$	283)
		Kr(111)	283)
	$CO + H_2$	$C(3 \times 3)$	301)
	H_2S	$(2 \times 2) - S$	36, 118, 197, 198)

Table 5.3 (continued)

Surface	Adsorbed gas	Surface structure	References
		$C(2 \times 2) - S$	36, 118, 197, 198, 293, 294, 303, 304, 340)
		$(2 \times 1) - S$	198)
		$C(2 \times 2) - H_2S$	304)
	H_2Se	$(2 \times 2) - Se$	197, 198)
		$C(2 \times 2) - Se$	142, 197, 198, 293, 294, 305, 340)
		$(2 \times 1) - Se$	198)
		$C(4 \times 2) - Se$	305)
	Te	$(2 \times 2) - Te$	197, 198, 306)
		$C(2 \times 2) - Te$	197, 198, 294, 305, 340)
		$(2 \times 1) - Te$	198)
		$C(4 \times 2) - Te$	305, 306)
	SO_2	$C(2 \times 2) - SO_2$	86)
		$(2 \times 2) - SO_2$	86)
NiO(100)	H_2	Adsorbed	307)
		Ni(100)	307)
	H_2S	$Ni(100) - C(2 \times 2) - S$	308)
	Cl_2	Disordered	309)
Pd(100)	CO	Disordered	70)
		$C(4 \times 2) - CO$	70)
		$C(2 \times 2) - CO$	210)
		$(2 \times 4)R45^\circ - CO$	71, 209, 210)
		Hexagonal overlayer	209, 210)
	Xe	Hexagonal overlayer	311)
Pt(100)	O_2	Not adsorbed	120, 312)
		Adsorbed	312, 135)
		$(2\sqrt{2} \times 2\sqrt{2})R45^\circ - O$	215, 313)
		$PtO_2(0001)$	215)
		$(5 \times 1) - O$	315)
		$(2 \times 1) - O$	315)
	CO	$C(4 \times 2) - CO$	28, 72, 73, 120, 314, 316)
		$(3\sqrt{2} \times \sqrt{2})R45^\circ - CO$	28, 72, 73, 316)
		$(\sqrt{2} \times \sqrt{5})R45^\circ - CO$	72, 73)
		$(2 \times 4) - CO$	10)
		$(1 \times 3) - CO$	10)
		$(1 \times 1) - CO$	120, 312, 314, 316)
		$C(2 \times 2) - CO$	312, 316)
	H_2	Adsorbed	312, 317)
		$(2 \times 2) - H$	72, 74)
		Not adsorbed	312)
	$CO + H_2$	$C(2 \times 2) - (CO + H_2)$	72, 74)
	NO	$(1 \times 1) - NO$	318)
		$C(4 \times 2) - NO$	319)
	N	Disordered	228)
	H_2S	$(2 \times 2) - S$	247, 320)
		$C(2 \times 2) - S$	247, 320, 321)
	S_2	$(2 \times 2) - S$	225, 226)

Table 5.3 (continued)

Surface	Adsorbed gas	Surface structure	References
		C(2 × 2) – S	225, 226)
Rh(100)	O_2	(2 × 2) – O	231)
		C(2 × 2) – O	231)
		C(2 × 8) – O	58)
	CO	C(2 × 2) – CO	231)
		Hexagonal overlayer	231)
		(4 × 1) – CO	58)
	CO_2	C(2 × 2) – CO	231)
		Hexagonal overlayer	231)
	H_2	Adsorbed	231)
	NO	C(2 × 2) – NO	231)
Si(100)	O_2	(1 × 1) – O	17, 18, 20)
		(111) facets	17, 18, 20)
	H_2	(1 × 1) – H	322, 323, 324)
		(2 × 1) – H	237)
	H	(1 × 1) – H	325)
		(2 × 1) – H	325)
	NH_3	(111) facets	238)
	I_2	(3 × 3) – 1	326)
Sr(100)	O_2	SrO(100)	327)
Ta(100)	O_2	(2 × 8/9) – O	328)
		C(3 × 1) – O	328)
		(4 × 1) – O	328)
	CO	C(3 × 1) – O	328)
	CO_2	C(3 × 1) – O	328)
	NO	C(3 × 1) – O	328)
	N_2	Adsorbed	328)
Th(100)	O_2	Disordered	329)
		ThO_2	329)
	CO	Disordered	329)
V(100)	O_2	(1 × 1) – O	65)
		(2 × 2) – O	65)
	H_2	Disordered	65)
W(100)	O_2	Disordered	330)
		(4 × 1) – O	66, 330-333, 336)
		(2 × 2) – O	330-334)
		(2 × 1) – O	66, 67, 330-336)
		(3 × 3) – O	331, 333, 335)
		C(2 × 2) – O	333)
		C(8 × 2) – O	333)
		(3 × 1) – O	333)
		(1 × 1) – O	333)
		(8 × 1) – O	333)
		(4 × 4) – O	333, 335)
		(110) facets	333)
	CO	Disordered	75)
		C(2 × 2) – CO	66, 75)
	H_2	C(2 × 2) – H	66, 78, 79, 337, 411)

Table 5.3 (continued)

Surface	Adsorbed gas	Surface structure	References
		C(2 × 2) – H	66, 78, 79, 337, 411)
		(2 × 5) – H	79)
		(4 × 1) – H	79)
		(1 × 1) – H	411)
	CO_2	Disordered	338)
		(2 × 1) – O	338)
		C(2 × 2) – CO	338)
	NO	(2 × 2) – NO	339)
		(4 × 1) – NO	339)
		(2 × 2) – O	339)
		(4 × 1) – O	339)
		(2 × 1) – O	339)
	N_2	C(2 × 2) – N	68, 82, 131)
	NH_3	Disordered	84)
		C(2 × 2) – NH_2	84)
		(1 × 1) – NH_2	84)
	N_2O	(1 × 1) – N_2O	143)
		(4 × 1) – N_2O	143)
	CO + N_2	(4 × 1) – (CO + N_2)	82)

Table 5.4. Surface structures on substrates with twofold rotational symmetry

Surface	Adsorbed gas	Surface structure	References
Ag(110)	O_2	(2 × 1) – O	146, 341, 342, 343, 344)
		(3 × 1) – O	146, 341, 342, 343)
		(4 × 1) – O	146, 341, 342)
		(5 × 1) – O	146, 341)
		(6 × 1) – O	146, 341)
		(7 × 1) – O	146)
	NO	Disordered	345)
	$C_2H_4Cl_2$	(2 × 1) – Cl	154)
		c(4 × 2) – Cl	154)
	Xe	Hexagonal overlayer	159)
Al(110)	O_2	(331) facets	123)
		(111) facets	122)
Au(110)	H_2S	(1 × 2) – S	251)
		c(4 × 2) – S	251)
C(110), diamond	O_2	Not adsorbed	164)
	N_2	Not adsorbed	164)
	NH_3	Not adsorbed	164)
	H_2S	Not adsorbed	164)
Cr(110)	O_2	(3 × 1) – O	140)
		(100) facets	140, 256)
		Cr_2O_3(0001)	140, 256)
Cu(110)	O_2	(2 × 1) – O	7, 8, 9, 45, 46, 246)
		c(6 × 2) – O	8, 9, 45, 46, 246)

Table 5.4 (continued)

Surface	Adsorbed gas	Surface structure	References
		$(5 \times 3) - O$	8, 115)
	CO	ordered 1D	26)
		$(2 \times 3) - CO$	26)
		$(2 \times 1) - CO$	255)
		Hexagonal overlayer	255)
	H_2	Not adsorbed	7)
	H_2O	Disordered	26)
	H_2S	$c(2 \times 3) - S$	35)
		Adsorbed	35)
	Xe	$c(2 \times 2) - Xe$	159)
		Hexagonal overlayer	159)
Cu/Ni(110)	O_2	$(2 \times 1) - O$	134)
	CO	$(2 \times 1) - CO$	134)
		$(2 \times 2) - CO$	134)
	H_2S	$c(2 \times 2) - S$	134)
Fe(110)	O_2	$c(2 \times 2) - O$	87, 88, 99)
		$c(3 \times 1) - O$	87, 88, 99)
		$(2 \times 8) - O$	98)
		FeO(111)	87, 88, 99, 269)
		$(2 \times 1) - O$	141)
	CO	$\begin{pmatrix} 3 & -2 \\ 0 & 4 \end{pmatrix} - CO$	346)
	N_2	$\begin{pmatrix} 3 & -2 \\ 0 & 4 \end{pmatrix} - N_2$	346)
	H_2	$(2 \times 1) - H$	177)
		$(3 \times 1) - H$	177)
		$(1 \times 1) - H$	177)
	H_2S	$(2 \times 4) - S$	114)
		$(1 \times 2) - S$	114)
Fe/Cr(110)	O_2	$Cr_2O_3(0001)$	280)
		Amorphous oxide	279)
Ge(110)	O_2	Disordered	17, 18)
		$(1 \times 1) - O$	17, 18)
	H_2S	$(10 \times 5) - S$	178)
Ir(110)	O_2	$(1 \times 2) - O$	347)
	CO	$(2 \times 2) - CO$	347, 348)
		$(4 \times 2) - CO$	348)
	H_2	Adsorbed	347)
	N_2	Not adsorbed	347)
$LaB_6(110)$	O_2	$(1 \times 1) - O$	349)
Mo(110)	O_2	$(2 \times 2) - O$	62, 63, 100)
		$(2 \times 1) - O$	62, 63, 100)
		$(1 \times 1) - O$	62, 63)
		Disordered	350)
	CO	$(1 \times 1) - CO$	62, 100)
		$c(2 \times 2) - CO$	94)
		Disordered	406)
	CO_2	Disordered	94)

Table 5.4 (continued)

Surface	Adsorbed gas	Surface structure	References
	H_2	Adsorbed	100)
	N_2	$(1 \times 1) - N$	62)
	H_2S	$(2 \times 2) - S$	351)
		$c(2 \times 2) - S$	351)
		$(1 \times 1) - S$	351)
		$c(1 \times 3) - S$	351)
		$c(1 \times 5) - S$	351)
		$(1 \times 3) - S$	351)
		$c(1 \times 7) - S$	351)
		$(1 \times 4) - S$	351)
		$(1 \times 5) - S$	351)
		$c(1 \times 11) - S$	351)
		$\begin{pmatrix} 2 & 2 \\ -1 & 1 \end{pmatrix} - S$	351)
Mo(211)	O_2	$(2 \times 1) - O$	105)
		$(1 \times 2) - O$	105)
		$(1 \times 3) - O$	105)
		$c(4 \times 2) - O$	105)
	CO	Disordered	105)
	H_2	$(1 \times 2) - H$	105)
	N_2	Not adsorbed	105)
Na(110)	O_2	NaO(111)	352)
Nb(110)	O_2	$(3 \times 1) - O$	101)
		NbO(111)	192)
		NbO(110)	192)
		NbO(220)	192)
		oxide	101)
	CO	Disordered	101)
		$(3 \times 1) - O$	101)
	H_2	$(1 \times 1) - H$	111)
Ni(110)	O_2	$(2 \times 1) - O$	2, 3, 51, 57, 83, 89, 91, 92, 99, 198, 353, 354, 355)
		$(3 \times 1) - O$	2, 51, 83, 89, 91, 92, 94, 198, 353, 354, 355
		$(5 \times 1) - O$	2, 89)
		$(9 \times 4) - O$	51, 354, 355)
		NiO(100)	6, 51, 83, 91, 198, 354, 355)
	CO	$(1 \times 1) - CO$	2, 94)
		Adsorbed	198)
		$c(2 \times 1) - CO$	353, 356, 359)
		$(2 \times 1) - CO$	356, 357, 358)
		$c(2 \times 2) - CO$	359)
		$(4 \times 2) - CO$	359)
	H_2	$(1 \times 2) - H$	59, 81, 94, 110, 198, 203, 353, 360)
	NO	$(2 \times 3) - N$	361)

Table 5.4 (continued)

Surface	Adsorbed gas	Surface structure	References
		(2 × 1) – O	361)
	H_2O	(2 × 1) – H_2O	110)
	H_2S	c(2 × 2) – S	36, 198, 205, 294)
		(3 × 2) – S	36)
	H_2Se	c(2 × 2) – Se	137)
	CO + O_2	(3 × 1) – (CO + O_2)	91)
Pd(110)	O_2	(1 × 3) – O	95)
		(1 × 2) – O	95)
		c(2 × 4) – O	95)
	CO	(5 × 2) – CO	95)
		(2 × 1) – CO	95, 209)
		(4 × 2) – CO	209)
		c(2 × 2) – CO	209)
	H_2	(1 × 2) – H	212)
Pt(110)	O_2	(2 × 1) – O	11, 363)
		(4 × 2) – O	11
		Adsorbed	362)
		c(2 × 2) – O	363)
		PtO(100)	363)
	CO	(1 × 1) – CO	139, 364)
		(2 × 1) – CO	366)
	C_3O_2	(1 × 1) – C_3O_2	365)
	NO	(1 × 1) – NO	222, 364)
	CO + NO	(1 × 1) – (CO + NO)	364)
	H_2S	c(2 × 6) – S	247, 367, 368)
		(2 × 3) – S	247, 367, 368)
		(4 × 3) – S	247, 367, 368)
		c(2 × 4) – S	247, 367, 368)
		(4 × 4) – S	247, 367)
Rh(110)	O_2	Disordered	96, 97)
		c(2 × 4) – O	96, 97)
		c(2 × 8) – O	96, 97)
		(2 × 2) – O	96, 97)
		(2 × 3) – O	96, 97)
		(1 × 2) – O	96, 97)
		(1 × 3) – O	96, 97)
	CO	(2 × 1) – CO	369)
		c(2 × 2) – C	369)
Ru(10$\bar{1}$0)	O_2	c(4 × 2) – O	370, 371)
		(2 × 1) – O	370, 371)
		c(2 × 6) – O	370)
		(7 × 1) – O	370)
		c(4 × 8) – O	370)
	CO	Disordered	371)
	H_2	Not adsorbed	371)
	N_2	Not adsorbed	371)
	NO	c(4 × 2) – (N + O)	370, 371)
		(2 × 1) – (N + O)	370, 371)
		(2 × 1) – O	371)

Table 5.4 (continued)

Surface	Adsorbed gas	Surface structure	References
		c(4 × 2) – O	371)
		c(2 × 6) – O	370)
		(7 × 1) – O	370)
		c(4 × 8) – O	370)
		(2 × 1) – N	371)
		c(4 × 2) – N	371)
Ru(101)	O_2	$\begin{pmatrix}1 & 1\\ 3 & 0\end{pmatrix} - O$	374)
		$\begin{pmatrix}2 & 1\\ 5 & 0\end{pmatrix} - O$	374)
		$\begin{pmatrix}4 & 1\\ 9 & 0\end{pmatrix} - O$	374)
	CO	$\begin{pmatrix}1 & 1\\ 3 & 0\end{pmatrix} - CO$	372)
		$\begin{pmatrix}0 & 1\\ 2 & 0\end{pmatrix} - C$	372)
	NO	Disordered	373)
Si(110)	H_2	(1 × 1) – H	375)
Si(311)	NH_3	Adsorbed	238)
Ta(100)	O_2	(3 × 1) – O	101, 102)
		Oxide	101, 102)
	CO	Disordered	101, 102)
		(3 × 1) – O	101, 102)
	H_2	(1 × 1) – H	102)
	N_2	Not adsorbed	101)
Ta(211)	O_2	(3 × 1) – O	101, 102)
		Oxide	101, 102)
	CO	Disordered	101, 102)
		(3 × 1) – O	102)
	H_2	(1 × 1) – H	102)
	N_2	Disordered	102)
		(311)facets	102)
TiO_2(100)	O_2	Disordered	376)
	H_2O	Disordered	376)
V(110)	O_2	(3 × 1) – O	101)
	CO	Disordered	101)
		(3 × 1) – O	101)
W(110)	O_2	(2 × 1) – O	57, 103, 377-387)
		c(2 × 2) – O	104)
		(2 × 2) – O	104, 387)
		(1 × 1) – O	104)
		c(14 × 7) – O	57, 103, 104)
		c(21 × 7) – O	104)
		c(48 × 16) – O	104)
		WO_3(100)	388)
		WO_3(111)	388)
	CO	Disordered	109)
		c(9 × 5) – CO	109)
		(1 × 1) – CO	379)

Table 5.4 (continued)

Surface	Adsorbed gas	Surface structure	References
		c(2 × 2) – CO	379)
		(2 × 7) – CO	389)
		c(4 × 1) – CO	389)
		(3 × 1) – CO	389)
		(4 × 1) – CO	389)
		(5 × 1) – CO	389, 390)
		(2 × 1) – (C + O)	389, 390)
		c(9 × 5) – (C + O)	389)
	$CO + O_2$	c(11 × 5) – ($CO + O_2$)	93)
	H_2	(2 × 1) – H	136)
	I_2	(2 × 2) – I	391)
		(2 × 1) – I	391)
W(211)	O_2	(2 × 1) – O	15, 106, 107, 108, 403, 404)
		(1 × 2) – O	15, 106, 404)
		(1 × 1) – O	106, 107, 403, 404)
		(1 × 3) – O	106)
		(1 × 4) – O	106, 404)
	CO	Disordered	108)
		c(6 × 4) – CO	108)
		(2 × 1) – CO	108)
		c(2 × 4) – CO	108)
	H_2	(1 × 1) – H	112)
	NH_3	c(4 × 2) – NH_2	113)
	$CO + O_2$	(1 × 1) – ($CO + O_2$)	108)
		(1 × 2) – ($CO + O_2$)	108)
W(210)	CO	(2 × 1) – CO	138)
		(1 × 1) – CO	138)
	N_2	(2 × 1) – N	131)
W(310)	N_2	(2 × 1) – N	131)
		c(2 × 2) – N	131)
ZnO(10$\bar{1}$0)	O_2	(1 × 1) – O	392)

Table 5.5. Miller indices, stepped surface designations and angles between the macroscopic surface and terrace planes for fcc crystals

Miller index	Stepped surface designation	Angle between the macroscopic surface and terrace (degrees)
(544)	(S) – [9(111) × (100)]	6.2
(755)	(S) – [6(111) × (100)]	9.5
(533)	(S) – [4(111) × (100)]	14.4
(211)	(S) – [3(111) × (100)]	19.5
(311)	(S) – [2(111) × (100)]	29.5
(311)	(S) – [2(100) × (111)]	25.2
(511)	(S) – [3(100) × (111)]	15.8
(711)	(S) – [4(100) × (111)]	11.4
(665)	(S) – [12(111) × (111)]	4.8

Table 5.5 (continued)

Miller index	Stepped surface designation	Angle between the macroscopic surface and terrace (degrees)
(997)	(S) – [9(111) × (111)]	6.5
(332)	(S) – [6(111) × (111)]	10.0
(221)	(S) – [4(111) × (111)]	15.8
(331)	(S) – [3(111) × (111)]	22.0
(331)	(S) – [2(110) × (111)]	13.3
(771)	(S) – [4(110) × (111)]	5.8
(610)	(S) – [6(100) × (100)]	9.5
(410)	(S) – [4(100) × (100)]	14.0
(310)	(S) – [3(100) × (100)]	18.4
(210)	(S) – [2(100) × (100)]	26.6
(210)	(S) – [2(110) × (100)]	18.4
(430)	(S) – [4(110) × (100)]	8.1
(10, 8, 7)	(S) – [7(111) × (310)]	8.5

Table 5.6. Surface structures on stepped substrates

Surface	Adsorbed gas	Surface structure	References
Ag(211)	Xe	Hexagonal overlayer	159)
Ag(331)	O_2	Disordered	393)
		Ag(110 – (2 × 1) – O	393)
	Cl_2	(6 × 1) – Cl	393)
Au(S) – [6(111) × (100)]	O_2	Oxide	161)
Cu(210)	O_2	(410), (530) facets	259)
Cu(211)	Xe	Hexagonal overlayer	156)
	Kr	Hexagonal overlayer	156)
Cu(311)	Xe	Hexagonal overlayer	394)
	CO	Adsorbed	394)
Cu(841)	O_2	(410), (100) facets	259)
Cu(S) – [3(100) × (100)]	CO	Not adsorbed	132)
	N_2	(1 × 2) – N	132)
Cu(S) – [4(100) × (100)]	O_2	(1 × 1) – O	132)
	CO	Not adsorbed	132)
	N_2	(1 × 3) – N	132)
Cu(S) – [4(100) × (111)]	H_2S	8(1d) – S	35)
Ir(S) – [6(111) × (100)]	O_2	(2 × 1) – O	182)
	CO	Disordered	182)
	H_2O	Not adsorbed	182)
	H_2	Adsorbed	187)
Ni(210)	O_2	facets	395)
	N_2	Ni(100) – $(6\sqrt{2} \times \sqrt{2})$ R45° – N	395)
		Ni(110) – (2 × 3) – N	395)
Pd(210)	CO	(1 × 1) – CO	209, 210)
		(1 × 2) – CO	209, 210)

Table 5.6 (continued)

Surface	Adsorbed gas	Surface structure	References
Pd(311)	CO	$(2 \times 1) - CO$	209)
		$3(1d) - CO$	209)
Pd(S) – [9(111) × (111)]	CO	$(\sqrt{3} \times \sqrt{3})R30° - CO$	209)
		Hexagonal overlayer	209)
Pt(S) – [4(111) × (100)]	H_2	facets	221)
Pt(S) – [6(111) × (100)]	O_2	$2(1d) - O$	120)
		$Pt(111) - (2 \times 2) - O$	215)
		$Pt(111) - (\sqrt{3} \times \sqrt{3})$ R30° – O	215)
		$Pt(111) - (\sqrt{79} \times \sqrt{79})$ R18° 7′ – O	215)
		$Pt(111) - (4 \times 2\sqrt{3})$ R30° – O	215)
		$Pt(111) - 3(1d) - O$	215)
	CO	Disordered	120)
	H_2	$2(1d) - H$	120, 221)
		Adsorbed	396)
		Pt(S) – [11(111) × 2(100)]	396)
Pt(S) – [9(111) × (100)]	H_2	$2(1d) - H$	221)
Pt(S) – [9(111) × (111)]	O_2	$(2 \times 2) - O$	397, 398, 399)
		Not adsorbed	120)
	CO	Disordered	120)
	H_2	$(2 \times 2) - H$	120)
		Adsorbed	400)
Pt(S) – [12(111) × (111)]	NO	(2×2) NO	401)
	NH_3	Disordered	401)
Re(S) – [14(0001) × (10$\bar{1}$1)]	CO	$(2 \times 2) - CO$	230)
		$(2 \times 1) - C$	230)
Rh(331)	O_2	$2(1d) - O$	402)
		$\begin{pmatrix} 1 & 2 \\ 2 & 0 \end{pmatrix} - O$	402)
		$\begin{pmatrix} 1 & 2 \\ 7 & -1 \end{pmatrix} - O$	402)
		facets	402)
	CO	$\begin{pmatrix} 1 & 2 \\ 5 & -1 \end{pmatrix} - CO$	402)
		$\begin{pmatrix} 1 & 2 \\ 2 & 0 \end{pmatrix} - CO$	402)
		Hexagonal overlayer	402)
	CO_2	$\begin{pmatrix} 1 & 2 \\ 5 & -1 \end{pmatrix} - CO$	402)
		$\begin{pmatrix} 1 & 2 \\ 2 & 0 \end{pmatrix} - CO$	402)
		Hexagonal overlayer	402)
	H_2	Adsorbed	402)
	NO	Disordered	402)
		$\begin{pmatrix} -1 & 1 \\ 3 & 0 \end{pmatrix}$	402)

Table 5.6 (continued)

Surface	Adsorbed gas	Surface structure	References
Rh(S) – [6(111) × (100)]	O_2	(2 × 2) – O	402)
		Rh(S) – [12(111) × 2(100)] – (2 × 2) – O	402)
		Rh(111) – (2 × 2) – O	402)
	CO	$(\sqrt{3} \times \sqrt{3})$ – R30° – CO	402)
		(2 × 2) – CO	402)
	CO_2	$(\sqrt{3} \times \sqrt{3})$ – R30° – CO	402)
		(2 × 2) – CO	402)
	H_2	Adsorbed	402)
	NO	(2 × 2) – NO	402)
W(S) – [6(110) × (1$\bar{1}$0)]	O_2	(2 × 1) – O	382)
W(S) – [8(110) × (112)]	O_2	(2 × 1) – O	382)
W(S) – [10(110) × (011)]	O_2	(2 × 1) – O	405)
W(S) – [12(110) × (1$\bar{1}$0)]	O_2	(2 × 1) – O	382)
W(S) – [16(110) × (112)]	O_2	(2 × 1) – O	382)
W(S) – [24(110) × (011)]	O_2	(2 × 1) – O	405)

Table 5.7. Surface structures formed by adsorption of organic compounds

Surface	Adsorbed gas	Surface structure	References
Ag(110)	HCN	Disordered	407)
	C_2N_2	Disordered	407)
Au(111)	C_2H_4	Not adsorbed	161)
	n-Heptane	Not adsorbed	161)
	Cyclohexene	Not adsorbed	161)
	Benzene	Not adsorbed	161)
	Naphthalene	Disordered	161)
Au(S) – [6(111) × (100)]	C_2H_4	Not adsorbed	161)
	n-Heptane	Not adsorbed	161)
	Cyclohexene	Not adsorbed	161)
	Benzene	Not adsorbed	161)
	Naphthalene	Disordered	161)
Cu(111)	C_2H_4	Not adsorbed	26)
	Fe-Phtalocyanine	Adsorbed	408)
	Cu-Phtalocyanine	Adsorbed	408)
	H-Phtalocyanine	Adsorbed	408)
	Glycine	(8 × 8)	409)
	L-Alanine	$(2\sqrt{13} \times 2\sqrt{13})R13°40'$	409)
	L-Tryptophan	$\begin{pmatrix} 7 & 1 \\ -2 & 4 \end{pmatrix}$	409)
	D-Tryptophan	$\begin{pmatrix} -8 & 1 \\ -2 & 4 \end{pmatrix}$	409)
Cu(100)	C_2H_4	(2 × 2)	26)
	Fe-Phtalocyanine	$\begin{pmatrix} 5 & -2 \\ 2 & 5 \end{pmatrix}$	408)

Table 5.7 (continued)

Surface	Adsorbed gas	Surface structure	References
	Cu-Phtalocyanine	$\begin{pmatrix} 5 & -2 \\ 2 & 5 \end{pmatrix}$	408)
	H-Phtalocyanine	$\begin{pmatrix} 5 & -2 \\ 2 & 5 \end{pmatrix}$	408)
	Glycine	(4×2)	409)
		$\begin{pmatrix} 8 & -4 \\ 4/5 & 8/5 \end{pmatrix}$	
	L-Alanine	$\begin{pmatrix} 2 & 1 \\ 2 & -1 \end{pmatrix}$	409)
	L-Tryptophan	(4×4)	409)
	D-Tryptophan	(4×4)	409)
Cu(110)	C_2H_4	ord. 1D	26)
Cu(S) – [3(100) × (100)]	CH_4	Not adsorbed	132)
	C_2H_4	Not adsorbed	132)
Cu(S) – [4(100) × (100)]	CH_4	Not adsorbed	132)
	C_2H_4	Not adsorbed	132)
Fe(100)	C_2H_4	$c(2 \times 2) - C$	274)
Ir(111)	C_2H_2	$(\sqrt{3} \times \sqrt{3})R30°$	187)
		$(9 \times 9) - C$	187)
	C_2H_4	$(\sqrt{3} \times \sqrt{3})R30°$	187)
		$(9 \times 9) - C$	187)
	Cyclohexane	Disordered	187)
		$(9 \times 9) - C$	187)
	Benzene	(3×3)	187)
		$(9 \times 9) - C$	187)
Ir(100)	C_2H_2	Disordered	281, 410)
		$c(2 \times 2) - C$	281, 410)
	C_2H_4	Disordered	410)
		$c(2 \times 2) - C$	410)
	Benzene	Disordered	410)
Ir(110)	C_2H_4	Disordered	347)
		$(1 \times 1) - C$	347)
	Benzene	Disordered	347)
		$(1 \times 1) - C$	
Ir(S) – [6(111) × (100)]	C_2H_2	(2×2)	187)
	C_2H_4	(2×2)	187)
	Cyclohexane	(2×2)	187)
	Benzene	Disordered	187)
Mo(100)	CH_4	$c(4 \times 4) - C$	286)
		$c(2 \times 2) - C$	286)
		$c(6\sqrt{2} \times 2\sqrt{2})R45° - C$	286)
		$(1 \times 1) - C$	286)
Ni(111)	CH_4	(2×2)	117)
		$(2 \times \sqrt{3})$	117)
	C_2H_2	(2×2)	412, 413)
	C_2H_4	(2×2)	29, 39, 412)

Table 5.7 (continued)

Surface	Adsorbed gas	Surface structure	References
	C_2H_6	(2×2)	39, 117)
		$(2 \times \sqrt{3})$	117)
		$(\sqrt{7} \times \sqrt{7})R19° - C$	29)
	Cyclohexane	$(2\sqrt{3} \times 2\sqrt{3})R30°$	414)
	Benzene	$(2\sqrt{3} \times 2\sqrt{3})R30°$	414, 415)
Ni(100)	CH_4	$c(2 \times 2)$	117)
		(2×2)	117)
	C_2H_2	$c(2 \times 2)$	416)
		(2×2)	416)
		$c(4 \times 2)$	417)
		$(2 \times 2) - C$	417)
	C_2H_4	$c(2 \times 2)$	88, 416)
		(2×2)	416)
		$c(4 \times 2)$	417)
		$(2 \times 2) - C$	417)
		$(\sqrt{7} \times \sqrt{7})R19° - C$	88)
	C_2H_6	$c(2 \times 2)$	117)
		(2×2)	117)
	Benzene	$c(4 \times 4)$	415)
Ni(110)	CH_4	(2×2)	117)
		(4×3)	117)
		$(4 \times 5) - C$	117, 418)
		$(2 \times 3) - C$	418)
	C_2H_4	$(2 \times 1) - C$	419, 420, 421)
		$(4 \times 5) - C$	419, 420)
		Graphite overlayer	420)
	C_2H_6	(2×2)	117)
	C_5H_{12}	(4×3)	422)
		(4×5)	422)
Pt(111)	C_2H_2	(2×1)	28)
		(2×2)	423, 424, 425)
	C_2H_4	(2×2)	40, 424, 425)
		(2×1)	28)
		$2(1d) - C$	221)
		Graphite overlayer	221, 426)
	n-Butane	$\begin{pmatrix} 2 & 1 \\ -1 & 2 \end{pmatrix}$	427)
		$\begin{pmatrix} 2 & 2 \\ -5 & 5 \end{pmatrix}$	427)
		$\begin{pmatrix} 3 & -2 \\ 2 & 5 \end{pmatrix}$	427)
	n-Pentane	$\begin{pmatrix} 2 & 1 \\ 0 & 6 \end{pmatrix}$	427)
	n-Hexane	$\begin{pmatrix} 2 & 1 \\ -1 & 3 \end{pmatrix}$	427)
	n-Heptane	$\begin{pmatrix} 2 & 1 \\ 0 & 8 \end{pmatrix}$	427)
		(2×2)	221)

Table 5.7 (continued)

Surface	Adsorbed gas	Surface structure	References
	n-Octane	$\begin{pmatrix} 2 & 1 \\ -1 & 4 \end{pmatrix}$	427)
	Cyclohexane	$\begin{pmatrix} 4 & -1 \\ 1 & 5 \end{pmatrix}$	427)
		Disordered	221)
		(2 × 2)	221)
		Graphite overlayer	221)
	Benzene	$\begin{pmatrix} -2 & 2 \\ 5 & 5 \end{pmatrix}$	221, 429)
		$\begin{pmatrix} 4 & -2 \\ 0 & 4 \end{pmatrix}$	428)
		$\begin{pmatrix} 4 & -2 \\ 0 & 5 \end{pmatrix}$	428)
		$\begin{pmatrix} -2 & 2 \\ 4 & 4 \end{pmatrix}$	429)
	Toluene	3(1d)	221, 430)
		(4 × 2)	430)
		Graphite overlayer	430)
	Naphthalene	(6 × 6)	224, 429)
		napthalene (001)	224)
	Pyridine	(2 × 2)	429)
	m-Xylene	2.6(1d)	430)
	Mesitylene	3.4(1d)	430)
	T-Butylbenzene	Disordered	430)
	N-Butylbenzene	Disordered	430)
	Aniline	3(1d)	430)
	Nitrobenzene	3(1d)	430)
	Cyanobenzene	3(1d)	430)
Pt(100)	C_2H_2	c(2 × 2)	28, 72, 321, 431, 432)
	C_2H_4	c(2 × 2)	28, 72, 313, 321, 431)
		Graphite overlayer	313, 426)
		(511), (311) facets	426)
	Benzene	Disordered	432)
		2(1d)	429)
	Naphthalene	(6 × 6)	429)
	Pyridine	(1 × 1)	429)
		c(2 × 2)	429)
	Toluene	3(1d)	430)
	M-xylene	3(1d)	430)
	Mesitylene	3 (1d)	430)
	T-Butylbenzene	Disordered	430)
	N-Butylbenzene	Disordered	430)
	Aniline	Disordered	430)
	Nitrobenzene	Disordered	430)
	Cyanobenzene	Disordered	430)
	C_2N_2	(1 × 1)	433)
Pt(110)	HCN	$\begin{pmatrix} 1 & 2/3 \\ -1 & 2/3 \end{pmatrix}$	434)

Table 5.7 (continued)

Surface	Adsorbed gas	Surface structure	References
		c(2 × 4)	434)
		(1 × 1)	434)
	C_2N_2	(1 × 1)	407, 435)
Pt(S) – [4(111) × (100)]	C_2H_4	Disordered	221)
		Graphite overlayer	221)
		Facets	221)
	Cyclohexane	Disordered	221)
		(4 × 2) – C	221)
	n-Heptane	(4 × 2)	221)
		(4 × 2) – C	221)
	Benzene	Disordered	221)
		Graphite overlayer	221)
		Facets	221)
	Toluene	Disordered	221)
		2(1d) – C	221)
Pt(S) – [6(111) × (100)]	C_2H_4	(2 × 2)	120, 221)
		$\begin{pmatrix} 3 & 2 \\ -2 & 5 \end{pmatrix}$ – C	221)
		$\begin{pmatrix} 6 & 1 \\ -1 & 7 \end{pmatrix}$ C	221)
		$(\sqrt{19} \times \sqrt{19})R23.4^\circ$ – C	426)
		Graphite overlayer	426)
	Cyclohexane	2(1d)	221)
	n-Heptane	(2 × 2)	221)
		$\begin{pmatrix} 1 & 1 \\ -1 & 2 \end{pmatrix}$	221)
		(9 × 9) – C	221)
	Benzene	3(1d)	221)
		(9 × 9) – C	221)
	Toluene	Disordered	221)
		(9 × 9) – C	221)
Pt(S) – [7(111) × (310)]	C_2H_4	Disordered	221)
		Graphite overlayer	221)
	Cyclohexane	Disordered	221)
	n-Heptane	Disordered	221)
	Benzene	Disordered	221)
	Toluene	Disordered	221)
		Graphite overlayer	221)
Pt(S) – [9(111) × (100)]	C_2H_4	Adsorbed	221)
	Cyclohexane	Disordered	221)
	n-Heptane	(2 × 2)	221)
		$\begin{pmatrix} 1 & 1 \\ -1 & 2 \end{pmatrix}$	221)
		(5 × 5) – C	221)

Table 5.7 (continued)

Surface	Adsorbed gas	Surface structure	References
		(2 × 2) – C	221)
		$\begin{pmatrix} 1 & 1 \\ -1 & 2 \end{pmatrix}$ – C	221)
		2(1d) – C	221)
	Benzene	Disordered	221)
		$\begin{pmatrix} 1 & 1 \\ -1 & 2 \end{pmatrix}$ – C	221)
		Graphite overlayer	221)
	Toluene	3(1d)	221)
		Graphite overlayer	221)
Pt(S) – 9(111) × (111)	C_2H_4	Disordered	120)
		Graphite overlayer	398, 399)
	N	Disordered	228)
Pt(S) – 5(100) × (111)	C_2H_4	Graphite overlayer	426)
		(511), (311) and (731) facets	426)
Re(0001)	C_2H_2	Disordered	436)
		$(2 \times \sqrt{3})R30^\circ$ – C	436)
	C_2H_4	Disordered	436)
		$(2 \times \sqrt{3})R30^\circ$ – C	436)
Rh(111)	C_2H_2	c(4 × 2)	231)
	C_2H_4	c(4 × 2)	231)
		(8 × 8) – C	231)
		(2 × 2)R30° – C	231)
		$(\sqrt{19} \times \sqrt{19})R23.4^\circ$ – C	231)
		$(2\sqrt{3} \times 2\sqrt{3})R30^\circ$ – C	231)
		(12 × 12) – C	231)
Rh(100)	C_2H_2	c(2 × 2)	231)
	C_2H_4	c(2 × 2)	231)
		c(2 × 2) – C	231)
		Graphite overlayer	231)
Rh(331)	C_2H_2	$\begin{pmatrix} -1 & 1 \\ 3 & 0 \end{pmatrix}$	402)
	C_2H_4	$\begin{pmatrix} -1 & 1 \\ 3 & 0 \end{pmatrix}$	402)
		Graphite overlayer	402)
Rh(S) – [6(111) × (100)]	C_2H_2	Disordered	402)
	C_2H_4	Disordered	402)
		(111), (100) facets	402)
Si(111)	C_2H_2	Disordered	437)
Si(311)	C_2H_2	c(1 × 1)	135)
		(2 × 1)	135)
		(3 × 1)	135)
	C_2H_4	c(1 × 1)	135)
		(2 × 1)	135)
		(3 × 1)	135)

Table 5.7 (continued)

Surface	Adsorbed gas	Surface structure	References
Ta(100)	C_2H_4	Adsorbed	328)
W(111)	CH_4	$(6 \times 6) - C$	41)
W(100)	CH_4	$(5 \times 1) - C$	41)
W(110)	C_2H_4	$(15 \times 3)\, R\alpha - C$	41)
		$(15 \times 12) R\alpha - C$	41)

References for Tables 5.2–5.7

1. Muller, K.: Zeits. Naturforschung, *20A*, 153 (1965)
2. MacRae, A. U.: Surface Sci. *1*, 319 (1964)
3. Germer, L. H., Schneiber, E. J. Hartman, C. D.: Phil. Mag. *5*, 222 (1960)
4. Park, R. L., Farnsworth, H. E.: Appl. Phys. Letters *3*, 167 (1963)
5. Edmonds, T., Pitkethly, R. C.: Surface Sci. *15*, 137 (1969)
6. MacRae, A. U.: Science *139*, 379 (1963)
7. Ertl, G.: Surface Sci. *6*, 208 (1967)
8. Takahashi, N. *et al.*: C. R. Acad. Sci. *269B*, 618 (1969)
9. Simmons, G. W., Mitchell, D. F., Lawless, K. R.: Surface Sci. *8*, 130 (1967)
10. Tucker, C. W., Jr.: Surface Sci, *2*, 516 (1964)
11. Tucker, C. W. Jr.: J. Appl. Phys. *35*, 1897 (1964)
12. Grant, J. T., Haas, T. W.: Surface Sci. *21*, 76 (1970)
13. Ellis, W. P.: J. Chem. Phys. *48*, 5695 (1968)
14. Ferrante, J., Barton, G. C.: NASA Tech. Note D-4735, 1968
15. Taylor, N. J.: Surface Sci. *2*, 544 (1964)
16. Marsh, J. B., Farnsworth, H. E.: Surface Sci. *1*, 3 (1964)
17. Schlier, R. E., Farnsworth, H. E.: J. Chem. Phys. *30*, 917 (1959)
18. Farnsworth, H. E., Schlier, R. E., George, T. H., Buerger, R. M.: J. Appl. Phys. *29*, 1150 (1958)
19. Lander, J. J., Morrison, J.: J. Appl. Phys. *34,* 1411 (1963)
20. Lander, J. J., Morrison, J.: J. Appl. Phys. *33*, 2089 (1962)
21. Rovida, G. *et al.*: Surface Sci. *14*, 93 (1969)
22. Adams, R. O.: The Structure and Chemistry of Solid Surfaces, ed. G. A. Somorjai, John Wiley and Sons, Inc., New York, 1969
23. Farnsworth, H. E., Zehner, D. M.: Surface Sci. *17*, 7 (1969)
24. Dooley, G. J., Haas, T. W.: Surface Sci. *19,* 1 (1970)
25. Campbell, B. D., Haque, C. A., Farnsworth, H. E.: The Structure and Chemistry of Solid Surfaces ed. G. A. Somorjai, John Wiley and Sons, Inc., New York, 1969
26. Ertl, G.: Surface Sci. *7*, 309 (1967)
27. Edmonds, T., Pitkethly, R. C.: Surface Sci. *17*, 450 (1969)
28. Morgan, A. E., Somorjai, G. A.: J. Chem. Phys. *51*, 3309 (1969)
29. Bertolini, J. C., Dalmai-Imelik, G.: Coll. Intern. CNRS, Paris, 7-11 July 1969
30. Lander, J. J., Morrison, J.: Surface Sci. *4*, 241 (1966)
31. Germer, L. H., MacRae, A. U.: J. Chem. Phys. *36*, 1555 (1962)
32. Van Bommel, A. J., Meyer, F.: Surface Sci. *8*, 381 (1967)
33. Lander, J. J., Morrison, J.: J. Chem. Phys. *37*, 729 (1962)
34. Heckingbottom, R.: The Structure and Chemistry of Solid Surfaces, ed. G. A. Somorjai, John Wiley and Sons, Inc., New York 1969.
35. Domange, J. L., Oudar, J.: Surface Sci. *11*, 124 (1968)
36. Perdereau, M., Oudar, J.: Surface Sci. *20*, 80 (1970)

37. Van Bommel, A. J., Meyer, F.: Surface Sci. *6*, 391 (1967)
38. Florio, J. V., Robertson, W. D.: Surface Sci. *18*, 398 (1969).
39. Bertolini, J. C., Dalmai-Imelik, G.: Rapport Inst. de Rech. sur la Catalyse, Villeurbanne, 1969
40. Smith, D. L., Merrill, R. P.: J. Chem. Phys. *52*, 5861 (1970)
41. Boudart, M., Ollis, D. F.: The Structure and Chemistry of Solid Surfaces, ed. G. A. Somorjai, John Wiley and Sons, Inc., New York, 1969
42. Jona, F.: J. Phys. Chem. Solids *28*, 2155 (1967)
43. Bedair, S. M., Hoffmann, F., Smith Jr., H. P.: J. Appl. Phys. *39*, 4026 (1968)
44. Farrell, H. H.: Ph.D. Dissertation, University of California, Berkeley, 1969
45. Jordan, L. K., Scheibner, E. J.: Surface Sci. *10*, 373 (1968)
46. Trepte, L., Menzel-Kopp, C., Mensel, E.: Surface Sci. *8*, 223 (1967)
47. Lee, R. N., Farnsworth, H. E.: Surface Sci. *3*, 461 (1965)
48. Grant, J. T.: Surface Sci. *18*, 228 (1969)
49. Schlier, R. E., Farnsworth, H. E.: J. Appl. Phys. *25*, 1333 (1954)
50. Farnsworth, H. E., Tuul, J.: J. Phys. Chem. Solids *9*, 48 (1958)
51. May, J. W., Germer, L. H.: Surface Sci. *11*, 443 (1968)
52. Schlier, R. E., Farnsworth, H. E.: Advances Catalysis *9*, 434 (1957)
53. Germer, L. H., Hartman, C. D.: J. Appl. Phys. *31*, 2085 (1960)
54. Farnsworth, H. E., Madden Jr., H. H.: J. Appl. Phys. *32*, 1933 (1961)
55. Park, R. L., Farnsworth, H. E.: J. Chem. Phys. 43, 2351 (1965)
56. Germer, L. H.: Advances Catalysis *13*, 191 (1962)
57. Germer, L. H., Stern, R., MacRae, A. U.: *Metal Surfaces* ASM, Metals Park, Ohio, 1963, p. 287
58. Tucker, C. W.: Jr., J. Appl. Phys. *37*, 3013 (1966)
59. Haque, C. A., Farnsworth, H. E.: Surface Sci. *1*, 378 (1964)
60. Pignocco, A. J., Pellissier, G. E.: J. Electrochem. Soc. *112*, 1188 (1965)
61. Kann, H. K. A., Feuerstein, S.: J. Chem. Phys. *50*, 3618 (1969)
62. Hayek, K., Farnsworth, H. E.: Surface Sci. *10*, 429 (1968)
63. Farnsworth, H. E. Hayek, K.: Suppl. Nuovo Cimento, *5*, 2 (1967)
64. Dooley, G. J., Haas, T. W.: J. Chem. Phys. *52*, 461 (1970)
65. Vijai, K. K., Packman, P. F.: J. Chem. Phys. *50*, 1343 (1969)
66. Estrup, P. J.: *The Structure and Chemistry of Solid Surfaces,* ed. G. A. Somorjai, John Wiley and Sons, Inc., New York, 1969
67. Anderson, J., Danforth, W. E.: J. Franklin Inst. *279*, 160 (1965)
68. Onchi, M., Farnsworth, H. E.: Surface Sci. *11*, 203 (1968)
69. Armstrong, R. A.: *The Structure and Chemistry of Solid Surfaces,* ed. G. A. Somorjai, John Wiley and Sons, Inc., New York, 1969
70. Tracy, J. C., Palmberg, P. W.: J. Chem. Phys. *51*, , 4852 (1969)
71. Park, R. L., Madden, H. H.: Surface Sci. *11*, 188 (1968)
72. Morgan, A. E., Somorjai, G. A.: Surface Sci. *12*, 405 (1968)
73. Burggraf, C., Mosser, A.: C. R. Acad. Sci. *268B*, 1167 (1969)
74. Morgan, A. E., Somorjai, G. A.: Trans. Am. Cryst. Assoc. *4*, 59 (1968)
75. Anderson, J., Estrup, P. J.: J. Chem. Phys. *46*, 563 (1967)
76. Onchi, M., Farnsworth, H. E.: Surface Sci. *13*, 425 (1969)
77. Dooley, G. J., Haas, T. W.: J. Chem. Phys. *52*, 993 (1970)
78. Tamm, P. W., Schmidt, L. D.: J. Chem. Phys. *51*, 5352 (1969)
79. Estrup, P. J., Anderson, J.: J. Chem. Phys. *45*, 2254 (1966)
80. Madden, H. H., Farnsworth, H. E.: J. Chem. Phys. *34*, 1186 (1961)
81. May, J. W., Germer, L. H.: *The Structure and Chemistry of Surfaces,* ed. G. A. Somorjai, John Wiley and Sons, Inc., New York, 1969
82. Estrup. P. J., Anderson, J.: J. Chem. Phys., *46*, 567 (1967)
83. Park, R. L., Farnsworth, H. E.: J. Appl. Phys. *35*, 2220 (1964)
84. Estrup, P. J., Anderson, J.: J. Chem. Phys. *49*, 523 (1968)
85. Margot, E. *et al.*: C. R. Acad. Sci. *270* C, 1261 (1970)

86. Tideswell, N. W., Ballingal, J. M.: J. Vac. Sci. Techn. *7*, 496 (1970)
87. Portele, F.: Zeits. Naturforschung *24 A*, 1268 (1969)
88. Dalmai-Imelik, G., Bertolini, J. C.: C. R. Acad. Sci. *270*, 1079 (1970)
89. Germer, L. H., MacRae, A. U.: J. Appl. Phys. *33*, 2923 (1962)
90. Germer, L. H., MacRae, A. U., Robert, A.: Welch Foundation Research, Bull. No. 11, 1961, p. 5
91. Park, R. L., Farnsworth, H. E.: J. Chem. Phys. *40*, 2354 (1964)
92. Germer, L. H., May, J. W., Szostak, R. J.: Surface Sci. *7*, 430 (1967)
93. May, J. W., Germer, L. H., Chang, C. C.: J. Chem. Phys. *45*, 2383 (1966)
94. Jackson, A. G., M. P. Hooker,: Surface Sci. *6*, 297 (1967)
95. Ertl, G., Rau, P.: Surface Sci. *15*, 443 (1969)
96. Tucker Jr., C. W.: J. Appl. Phys. *38*, 2696 (1967)
97. Tucker Jr., C. W.: J. Appl. Phys. *37*, 4147 (1966)
98. Pignocco, A. J., Pellisier, G. E.: Surface Sci. *7*, 261 (1967)
99. Moliere, K., Portele, F.: *The Structure and Chemistry of Solid Surfaces,* ed. G. A. Somorjai, John Wiley and Sons, Inc., N. Y. 1969
100. Haas, T. W., Jackson, A. G.; J. Chem. Phys. *44*, 2921 (1966)
101. Haas, T. W., Jackson, A. G., Hooker, M. P.: J. Chem. Phys. *46*, 3025, (1967)
102. Haas, T. W.: *The Structure and Chemistry of Solid Surfaces,* ed. G. A. Somorjai, John Wiley and Sons, Inc., N. Y. 1969
103. Germer, L. H.: Physics Today, July 1964, p. 19
104. Germer, L. H., May, J. W.: Surface Sci. *4*, 452 (1966)
105. Dooley, G. J., Haas, T. W.: J. Vac. Sci. Techn. *7*, 49 (1970)
106. Chang, C. C., Germer, L. H.: Surface Sci. *8*, 115 (1967)
107. Tracy, T. C., Blakely, J. M.: *The Structure and Chemistry of Solid Surfaces*, ed. G. A. Somorjai, John Wiley and Sons, Inc., N. Y., 1969
108. Chang, C. C.: J. Electrochem. Soc. *115*, 354 (1968)
109. May, J. W., Germer, L. H.: J. Chem. Phys. *44*, 2895 (1966)
110. Germer, L. H., MacRae, A. U.: Proc. Natl. Acad. Sci. U.S. *48*, 997 (1962)
111. Haas, T. W.: J. Appl. Phys. *39*, 5854 (1968).
112. Adams, D. L. *et al.*: Surface Sci. *22*, 45 (1970)
113. May, J. W., Szostak, R. J., Germer, L. H.: Surface Sci. *15*, 37 (1969)
114. Buckley, D. H.: NASA Techn. Note D-5689, 1970
115. Marklund, I., Andersson, S., Martinsson, J.: Arkiv for Fysik *37*, 127 (1968)
116. Legare, P., Marie, G.: J. Chem.Phys. Physsichim. Biol. *68* (7-8), 1206 (1971)
117. Marie, G., Anderson, J. R., Johnson, B. B.: Proc. Roy. Soc. Lond. A, *320*, 227 (1970)
118. Edmonds, T., McCarrol, J. J., Pitkethly, R. C.: J. Vac. Sci. Tech. *8* (1), 68 (1971)
119. Okado, K., Halsushika, T., Tomita, H., Motov, S., Takalashi, N.: Shinku *13* (11), 371 (1970)
120. Lang, B., Joyner, R. W., Somorjai, G. A.: Surf. Sci. *30*, 454 (1972)
121. Henzler, M., Töpler, J.: Surf. Sci. *40*, 388 (1973)
122. Van Hove, H., Leysen, R.: Phys. Status Solidi A, *9* (1), 361 (1972)
123. Bedair, S. M., Smith, H. P.: Jr., J. Appl. Phys. *42*, 3616 (1971)
124. Grant, J. T.: Surface Sci. *25*, 451 (1971)
125. Joyner, R. W., McKee, C. S., Roberts, M. W.: Surface Sci. *26*, 303 (1971)
126. Tracy, J. C.: J. Chem. Phys. *56* (6), 2748 (1971)
127. Chesters, M. A., Pritchard, J.: Surface Sci. *28*, 460 (1971)
128. Joyner, R. W., McKee, C. S., Roberts, M. W.: Surface Sci. *27*, 279 (1971)
129. Tracy, J. C.: J. Chem. Phys. *56* (6), 2736 (1971)
130. Tabor, D., Wilson, J. M.: J. Cryst. Growth *9*, 60 (1971)
131. Adams, D. L., Germer, L. H.: Surface Sci. *27*, 21 (1971)
132. Perdereau, J., Rhead, G. E.: Surface Sci. *24*, 555 (1971)
133. Palmberg, P. W.: Surface Sci. *25*, 104 (1971)
134. Ertl, G., Küppers, J.: Surface Sci. *24*, 104 (1971)

135. Heckingbottom, R., Wood, P. R.: Surface Sci. *23*, 437 (1970)
136. Matysik, K. J.: Surface Sci. *29*, 324 (1972)
137. Becker, G. E., Hagstrum, H. D.: Surface Sci. *30*, 505 (1972)
138. Adams, D. L., Germer, L. H.: Surface Sci. *32*, 205 (1972)
139. Bonzel, H. P., Ku, R.: Surface Sci. *33*, 91 (1972)
140. Michel, P., Jardin, Ch.: Surface Sci. *36*, 478 (1973)
141. Melmed, A., Carroll, J. J.: J. Vac. Sci. Technol., *10*, 164 (1973)
142. Hagstrum, H. D., Becker, G. E.: Phys. Rev. Lett. *22*, 1054 (1969); J. Chem. Phys. *54*, 1015 (1971)
143. Weinberg, W. H., Merrill, R. P.: Surface Sci. *32*, 317 (1972)
144. Sewell, P. B., Mitchell, D. F., Cohen, M.: Surface Sci. *33*, 535 (1972)
145. Forstmann, F., Berndt, W., Büttner, P.: Phys. Rev. Lett. *30*, 17 (1973)
146. Engelhardt, H. A., Menzel, D.: Surface Sci. *57* (1976) 591
147. Albers, H., Van der Wal, W. J. J., Bootsma, G. A.: Surface Sci. *68* (1977) 47
148. Rovida, G., Pratesi, F., Maglietta, M., Ferroni, E.: Surface Sci. *43* (1974) 230
149. Berndt, W.: Proc 2nd International Conference on Solid Surfaces (1974) 653
150. Forstmann, F.: Proc 2nd International Conference on Solid Surfaces (1974) 657
151. Goddard, P. J., Lambert, R. M.: Surface Sci *67* (1977) 180
152. Tu. Y., Blakely, J. M.: Journal of Vacuum Science Technology *15* (1978) 563
153. Rovida, G., Pratesi, F., Maglietta, M., Ferroni, E.: Proc 2nd International Conference on Solid Surfaces (1974) 117
154. Rovida, G., Pratesi, F.: Surface Sci. *51* (1975) 270
155. Goddard, P. J., Schwaha, K., Lambert, R. M.: Surface Sci. *71* (1978) 351
156. Roberts, R. H., Pritchard, J.: Surface Sci. *54* (1976) 687
157. Stoner, N., Van Hove, M. A., Tong, S. Y., Webb, M. B.: Phys. Rev. Letters *40* (1978) 243
158. McElhiney, G., Papp, H., Pritchard, J.: Surface Sci. *54* (1976) 617
159. Chesters, M. A., Hussain, M., Pritchard, J.: Surface Sci. *35* (1973) 161
160. Cohen, P. I., Unguris, J., Webb, M. B.: Surface Sci. *58*,((1976) 429
161. Chesters, M. A., Somorjai, G. A.: Surface Sci. *52* (1975) 21
162. Zehner, D. M., Wendelken, J. F.: Proc. 7th Internal Vacuum Congress and 3rd International Conference on Solid Surfaces (1977) 51
163. Goddard, P. J., West, J., Lambert, R. M.: Surface Sci. *71* (1978) 447
164. Lurie, P. G., Wilson, J. M.: Surface Sci. *65* (1977) 453
165. Suzanne, J., Coulomb, J. P., Bienfait, M.: Surface Sci. *40* (1973) 414
166. Chinn, M. D., Fain, S. C. Jr.: Journal of Vacuum Science Technology 14 (1977) 314.
167. Kramer, H. M., Suzanne, J.: Surface Sci. *54* (1976) 659
168. Bridge, M. E., Comrie, C. M., Lambert, R. M.: Surface Sci. *67* (1977) 393
169. Jardin, C., Michel, P.: Surface Sci. *71* (1978) 575
170. Habraken, F. H. P. M., Kieffer, E. P., Bootsma, G. A.: Proc 7th International Vacuum Congress and 3rd International Conference on Solid Surfaces (1977) 877
171. McDonnell, L., Woodruff, D. P.: Surface Sci. *46* (1974) 505
172. Kessler, J., Thieme, F.: Surface Sci. *67* (1977) 405
173. Benndorf, C., Gressman, K. H., Thieme, F.: Surface Sci. *61* (1976) 646
174. Chinn, M. D., Fain, S. C. Jr.: Physical Review Letters *39* (1977) 146
175. Nakanishi, S., Horiguchi, T.: Proc. 7th International Vacuum Congress and 3rd International Conference on Solid Surfaces (1977) A2727
176. Grunze, M., Bozso, F., Ertl, G., Weiss, M.: Applications of Surface Science *1* (1978) 241
177. Bozso, F., Ertl, G., Grunze, M., Weiss, M.: Applications of Surface Science *1* (1978) 103
178. Olshanetsky, B. Z., Repinsky, S. M., Shklyaev, A. A.: Surface Sci. *64* (1977) 224
179. Sinharoy, S., Henzler, M.: Surface Sci. *51* (1975) 75
180. Ivanov, V. P., Boreskov, G. K., Savchenko, V. I., Egelhoff, W. F. Jr., Weinberg, W. H.: Journal of Catalysis *48* (1977) 269

181. Conrad, H., Küppers, J., Nitschke, F., Plagge, A.: Surface Sci. *69* (1977) 668
182. Hagen, D. I., Nieuwenhuys, B. E., Rovida, G., Somorjai, G. A.: Surface Sci. *57* (1976) 632
183. Küppers, J., Plagge, A.: Journal of Vacuum Science Technology *13* (1976) 259
184. Ivanov, V. P., Boreskov, G. K., Savchenko, V. I., Egelhoff, W. F. Jr., Weinberg, W. H.: Surface Sci. *61* (1976) 207
185. Comrie, C. M., Weinberg, W. H.: Journal of Vacuum Science Technology *13* (1976) 264
186. Comrie, C. M., Weinberg, W. H.: J. Chem. Phys. *64* (1976) 250
187. Nieuwenhuys, B. E., Hagen, D. I., Rovida, G., Somorjai, G. A.: Surface Sci. *59* (1976) 155
188. Kanski, J., Rhodin, T. N.: Surface Sci. *65* (1977) 63
189. Clark, L. J.: Proc 7th International Vacuum Congress and 3rd International Conference on Solid Surfaces (1977) A2725
190. Kennett, H. M., Lee, A. E.: Surface Sci. *48* (1975) 606
191. Wilson, J. M.: Surface Sci. *59* (1976) 315
192. Pantel, R., Bujor, M., Bardolle, J.: Surface Sci. *62* (1977) 739
193. Conrad, H., Ertl, G., Küppers, J., Latta, E. E.: Surface Sci. *50* (1975) 296
194. Holloway, P. H., Hudson, J. B.: Surface Sci. *43* (1974) 141
195. Conrad, H., Ertl, G., Küppers, J., Latta, E. E.: Surface Sci. *57* (1976) 475
196. Erley, W., Besocke, K., Wagner, H.: J. Chem. Phys. *66* (1977) 5269
197. Marcus, P. M., Demuth, J. E., Jepsen, D. W.: Surface Sci. *53* (1975) 501
198. Demuth, J. E., Rhodin, T. N.: Surface Sci. *45* (1974) 249
199. Ertl, G.: Surface Sci. *47* (1975) 86
200. Christmann, K., Schober, O., Ertl, G.: J. Chem. Phys. *60* (1974) 4719
201. Van Hove, M. A., Ertl, G., Weinberg, W. H., Christmann, K., Behm, H. J.: Proc 7th International Conference on Solid Surfaces (1977) 2415
202. Conrad, H., Ertl, G., Küppers, J., Latta, E. E.: Surface Sci. *58* (1976) 578
203. Christmann, K., Schober, O., Ertl, G., Neumann, M.: J. Chem. Phys. *60* (1974) 4528
204. Casalone, G., Cattania, M. G., Simonetta, M., Tescari, M.: Surface Sci. *72* (1978) 739
205. Demuth, J. E., Jepsen, D. W., Marcus, P. M.: Phys. Rev. Lett. *32* (1974) 1182
206. Erley, W., Wagner, H.: Surface Sci. *66* (1977) 371
207. Conrad, H., Ertl, G., Küppers, J., Latta, E. E.: Surface Sci. *65* (1977) 245
208. Conrad, H., Ertl, G., Küppers, J., Latta, E. E.: Surface Sci. *65* (1977) 235
209. Conrad, H., Ertl, G., Koch, J., Latta, E. E.: Surface Sci. *43* (1974) 462
210. Bradshaw, A. M., Hoffman, F. M.: Surface Sci. *72* (1978) 513
211. Christmann, K., Ertl, G., Schober, O.: Surface Sci. *40* (1973) 61
212. Conrad, H., Ertl, G., Latta, E. E.; Surface Sci. *41* (1974) 435
213. Bonzel, H. P., Ku, R.: Surface Sci. *40* (1973) 85
214. Carrière, B., Deville, J. P., Maire, G., Légaré, P.: Surface Sci. *58* (1976) 578
215. Légaré, P., Maire, G., Carière, B., Deville, J. P.: Surface Sci. *68* (1977) 348
216. Joebstl, J. A.: Journal of Vacuum Science Technology *12* (1975) 347
217. Weinberg, W. H., Monroe, D. R., Lampton, V., Merrill, R. P.: Journal of Vacuum Science Technology *14* (1977) 444
218. Ertl, G., Neumann, M., Streit, K. M.: Surface Sci. *64* (1977) 393
219. Bernasek, S. L., Somorjai, G. A.: J. Chem. Phys. *60* (1974) 4552
220. Christmann, K., Ertl, G., Pignet, T.: Surface Sci. *54* (1976) 365
221. Baron, K., Blakely, D. W., Somorjai, G. A.: Surface Sci. *41* (1974) 45
222. Comrie, C. M., Weinberg, W. H., Lambert, R. M.: Surface Sci. *57* (1976) 619
223. Firment, L. E., Somorjai, G. A.: J. Chem. Phys. *63* (1975) 1037
224. Firment, L. E., Somorjai, G. A.: Surface Sci. *55* (1976) 413
225. Heegemann, W., Bechtold, E., Hayek, K.: Proc 2nd International Conference on Solid Surfaces (1974) 185
226. Heegemann, W., Meister, K. H., Bechtold, E., Hayek, K.: Surface Sci. *49* (1975) 161
227. Berthier, Y., Perdereau, M., Oudar, J.: Surface Sci. *44* (1974) 281

228. Schwaka, K., Bechtold, E.: Surface Sci. *66* (1977) 383
229. Gorodetsky, D. A., Knysh, A. N.: Surface Sci. *40* (1973) 651
230. Housley, M., Ducros, R., Piquard, G., Cassuto, A.: Surface Sci. *68* (1977) 277
231. Castner, D. G., Sexton, B. A., Somorjai, G. A.: Surface Sci. *71* (1978) 519.
232. Madey, T. E., Engelhardt, H. A., Menzel, D.: Surface Sci. *48* (1975) 304
233. Madey, T. E., Menzel, D.: Proc 2nd International Conference on Solid Surfaces (1974) 229
234. Danielson, L. R., Dresser, M. J., Donaldson, E. E., Dickinson, J. T.: Surface Sci. *71* (1978) 599
235. Danielson, L. R., Dresser, M. J., Donaldson, E. E., Sandstrom, D. R.: Surface Sci. *71* (1978) 615
236. Pandey, K. C., Sakurai, T., Hagstrum, H. D.: Phys. Rev. B. *16* (1977) 3648
237. Ibach, H., Rowe, J. E.: Surface Sci. *43* (1974) 481.
238. Heckingbottom, R., Wood, P. R.: Surface Sci. *36* (1973) 594
239. Van Bommel, A. J., Crombeen, J. E.: Surface Sci. *36* (1973) 773
240. Shih, H. D., Jona, F., Jepsen, D. W., Marcus, P. M.: Journal of Vacuum Science Technology *15* (1978) 596
241. Shih, H. D., Jona, F., Jepsen, D. W., Marcus, P. M.: Physical Review Letters *36* (1976) 798
242. Shih, H. D., Jona, F., Jepsen, D. W., Marcus, P. M.: Surface Sci. *60* (1976) 445
243. Bastasz, R., Colmenares, C. A., Smith, R. L., Somorjai, G. A., Surface Sci. *67* (1977) 45
244. Madey, T. E., Czyzewski, J. J., Yates, J. T. Jr.: Surface Sci. *57* (1976) 580
245. Unertl, W. N., Blakely, J. M.: Surface Sci. *69* (1977) 23
246. Oustry, A., Lafourcade, L., Escaut, A.: Surface Sci. *40* (1973) 545
247. Berthier, Y., Perdereau, M., Oudar, J.: Surface Sci. *36* (1973) 225
248. Fuggle, J. C., Umbach, E., Feulner, P., Menzel, D.: Surface Sci. *64* (1977) 69
249. Zanazzi, E., Jona, F., Jepsen, D. W., Marcus, P. M.: Phys. Rev. B *14* (1976) 432
250. Ignatiev, A., Jona, F., Jepsen, D. W., Marcus, P. M.: Surface Sci. *40* (1973) 439
251. Kostelitz, M., Domange, J. L., Oudar, J.: Surface Sci. *34* (1973) 431
252. McElhiney, G., Pritchard, J.: Surface Sci. *60* (1976) 397
253. Maglietta, M., Rovida, G.: Surface Sci. *71* (1978) 495
254. Rovida, G., Maglietta, M.: Proc. 7th International Vacuum Congress and 3rd International Conference on Solid Surfaces (1977) 963
255. Horn, K., Hussain, M., Pritchard, J.: Surface Sci. *63* (1977) 244
256. Ekelund, S., Leygraf, C.: Surface Sci. *40* (1973) 179
257. McDonnell, L., Woodruff, D. P., Mitchell, K. A. R.: Surface Sci. *45* (1974) 1
258. Tibbetts, G. G., Burkstrand, J. M., Tracy, J. C.: Phys. Rev. B. *15* (1977) 3652
259. Legrand-Bonnyns, E., Ponslet, A.: Surface Sci. *53* (1975) 675
260. Tibbetts, G. G., Burkstrand, J. M., Tracy, J. C.: Journal of Vacuum Science Technology *13* (1976) 362
261. McRae, E. G., Caldwell, C. W.: Surface Sci. *57* (1976) 77
262. Noonan, J. R., Zehner, D. M., Jenkins, L. H.: Surface Sci. *69* (1977) 731
263. Hoffmann, P., Unwin, R., Wyrobisch, W., Bradshaw, A. M.: Surface Sci. *72* (1978) 635
264. Gerhardt, U., Franz-Moller, G.; Proc. 7th International Vacuum Congress and 3rd International Conference on Solid Surfaces (1977) 897
265. Brundle, C. R., Wandelt, K.: Proc. 7th International Vacuum Congress and 3rd International Conference on Solid Surfaces (1977) 1171
266. Burkstrand, J. M., Kleiman, G. G., Tibbetts, G. G., Tracy, J. C.: Journal of Vacuum Science Technology *13* (1976) 291
267. Salwen, A., Rundgren, J.: Surface Sci. *53* (1975) 523
268. Legg, K. O., Jona, F., Jepsen, D. W., Marcus, P. M.: Physical Review B *16* (1977) 5271
269. Leygraf, C., Ekelund, S.: Surface Sci. *40* (1973) 609
270. Simmons, G. W., Dwyer, D. J.: Surface Sci. *48* (1975) 373
271. Brucker, C. F., Rhodin, T. N.: Surface Sci. *57* (1976) 523
272. Horiguchi, T., Nakanishi, S.: Proc. 2nd International Conference on Solid Surfaces (1974) 89
273. Watanabe, M., Miyarmura, M. Matsudaira, T., Onchi, M.: Proc. 2nd International Conference on Solid Surfaces (1974) 501

274. Brucker, C., Rhodin, T.: Journal of Catalysis *47* (1977) 214
275. Jona, F., Legg, K. O., Shih, H. D., Jepsen, D. W., Marcus, P. M.: Physical Review Letters *40* (1978) 1466
276. Matsudaira, T., Watanabe, M., Onchi, M.: Proc. 2nd International Conference on Solid Surfaces (1974) 181
277. Legg, K. O., Jona, F., Jepsen, D. W., Marcus, P. M.: Surface Sci. *66* (1977) 25
287. Dwyer, D. J., Simmons, G. W.: Surface Sci. *64* (1977) 617
279. Leygraf, C., Hultquist, G., Ekelund, S.: Surface Sci. *51* (1975) 409
280. Leygraf, C., Hultquist, G.: Surface Sci. *61* (1976) 69
281. Rhodin, T. N., Broden, G.: Surface Sci. *60* (1976) 466
282. Brodén, G., Rhodin, T. N.: S. S. Comm *18* (1976) 105
283. Ignatiev, A., Rhodin, T. N., Tong, S. Y.: Surface Sci. *42* (1974) 37
284. Riwan, R., Guillot, C., Paigne, J.: Surface Sci. *47* (1975) 183
285. Lecante, J., Riwan, R., Guillot, G.: Surface Sci. *35* (1973) 271
286. Guillot, C., Riwan, R., Lecante, J.: Surface Sci. *59* (1976) 581
287. Ignatiev, A., Jona, F., Jepsen, D. W., Marcus, P. M.: Surface Sci. *49* (1975) 189
288. Wilson, J. M.: Surface Sci. *53* (1975) 330
289. Glachant, A., Coulomb, J. P., Biberian, J. P.: Surface Sci. *59* (1976) 619
290. Farrell, H. H., Strongin, M.: Surface Sci. *38* (1973) 18
291. Brongersma, H. H., Theeten, J. B.: Surface Sci. *54* (1976) 519
292. Murata, Y., Ohtani, S., Terada, K.: Proc 2nd International Conference on Solid Surfaces (1974) 837
293. Demuth, J. E., Jepsen, D. W., Marcus, P. M.: Journal of Vacuum Science Technology *11* (1974) 190
294. Rhodin, T. N., Demuth, J. E.: Proc. 2nd International Conference on Solid Surfaces (1974) 167
295. Andersson, S., Kasemo, B., Pendry, J. B., Van Hove, M. A.: Phys. Rev. Lett. *31* (1973) 595
296. McRae, E. G., Caldwell, C. W.: Surface Sci. *57* (1976) 63
297. Holloway, P. H., Hudson, J. B.: Surface Sci. *43* (1974) 123
298. Dalmai-Imelik, G., Bertolini, J. C., Rousseau, J.: Surface Sci. *63* (1977) 67
299. Mitchell, D. F., Sewell, P. B., Cohen, M.: Surface Sci. *61* (1976) 355
300. Andersson, S., Pendry, J. B.: Surface Sci. *71* (1978) 75
301. Andersson, S.: Proc. 3rd International Vacuum Congress and 7th International Conference on Solid Surfaces (1977) 1019
302. Horn, K., Bradshaw, A. M., Jacobi, K.: Surface Sci. *72* (1978) 719
303. Demuth, J. E., Jepsen, D. W., Marcus, P. M.: Surface Sci. *45* (1974) 733
304. Matsudaira, T., Nishijima, M., Onchi, M.: Surface Sci. *61* (1976) 651
305. Froitzheim, H., Hagstrum, H. D.: Journal of Vacuum Science Technology *15* (1978) 485
306. Becker, G. E., Hagstrum, H. D.: Journal of Vacuum Science Technology *11* (1974) 234
307. Rickard, J. M., Perdereau, M., Dufour, L. G.: Proc. 7th International Vacuum Congress and 3rd International Conference on Solid Surfaces (1977) 847
308. Steinbrunn, A., Dumas, P., Colson, J. C.: Surface Sci. *74* (1978) 201
309. Netzer, F. P., Prutton, M.: Surface Sci. *52* (1975) 505
310. Pagageorgopoulos, C. A., Chen, J. M.: Surface Sci. *52* (1975) 40
311. Palmberg, P. W.: Surface Sci. *25* (1971) 104
312. Helms, C. R., Bonzel, H. P., Kelemen, S.: J. Chem. Phys. *65* (1976) 1773
313. Lang, B., Légaré, P., Maire, G.: Surface Sci. *47* (1975) 89
314. Kneringer, G., Netzer, F. P.: Surface Sci. *49* (1975) 125
315. Pirug, G., Brodén, G., Bonzel, H. P.: Proc. 7th International Vacuum Congress and 3rd International Conference on Solid Surfaces (1977) 907
315. Brodén, G., Pirug, G., Bonzel, H. P.: Surface Sci. *72* (1978) 45
317. Netzer, F. P., Kneringer, G.: Surface Sci. *51* (1975) 526
318. Bonzel, H. P., Pirug, G.: Surface Sci. *62* (1977) 45
319. Bonzel, H. P., Brodén, G., Pirug, G.: Journal of Catalysis *53* (1978) 96

320. Fischer, T. E., Kelemen, S. R.: Surface Sci. *69* (1977) 1
321. Fischer, T. E., Kelemen, S. R.: Journal of Vacuum Science Technology *15* (1978) 607
322. White, S. J., Woodruff, D. P.: Surface Sci. *63* (1977) 254
323. White, S. J., Woodruff, D. P., Holland, B. W., Zimmer, R. S.: Surface Sci. *74* (1978) 34
324. White, S. J., Woodruff, D. P., Holland, B. W., Zimmer, R. S.: Surface Sci. *68* (1977) 457
325. Sakurai, T., Hagstrum, H. D.: Phys. Rev. B *14* (1976) 1593
326. Lander, J. J., Morrison, J.: J. Chem. Phys. *37* (1962) 729
327. Janssen, A. P., Schoonmaker, R. C.: Surface Sci. *55* (1976) 109
328. Chesters, M. A., Hopkins, B. J., Leggett, M. R.: Surface Sci. *43* (1974) 1
329. Taylor, T. N., Colmenares, C. A., Smith, R. L., Somorjai, G. A., Surface Sci. *54* (1976) 317
330. Hopkins, B. J., Watt, G. D., Jones, A. R.: Surface Sci. *52* (1975) 715
331. Papageorgopoulous, C. A., Chen, J. M.: Surface Sci. *39* (1973) 313
332. Bradshaw, A. M., Menzel, D., Steinkilberg, M.: Proc. 2nd International Conference on Solid Surfaces (1974) 841
333. Bauer, E., Poppa, H., Viswanath, Y.: Surface Sci. *58* (1976) 578
334. Prigge, S., Niehus, H. Bauer, E.: Surface Sci. *65* (1977) 141
335. Desplat, J. L.: Proc. 2nd International Conference on Solid Surfaces (1974) 177
336. Luscher, P. E., Propst, F. M.: Journal of Vacuum Science Technology *14* (1977) 400
337. Jaeger, R., Menzel, D.: Surface Sci. *63* (1977) 232
338. Hopkins, B. J., Jones, A. R., Winton, R. I.: Surface Sci. *57* (1976) 266
339. Usami, S., Nakagima, T.: Proc. 2nd International Conference on Solid Surfaces (1974) 237
340. Demuth, J. E., Jepsen, D. W., Marcus, P. M.: Phys. Rev. Lett *31* (1973) 540
341. Engelhardt, H. A., Bradshaw, A. M., Menzel, D.: Surface Sci. *40* (1973) 410
342. Rovida, G., Pratesi, F.: Surface Sci. *52* (1975) 542
343. Heiland, W., Iberl, F., Taglauer, E., Menzel, D.: Surface Sci. *53* (1975) 383
344. Zanazzi, E., Maglietta, M., Bardi, U., Jona, F., Jepsen, D. W., Marcus, P. M.: Proc. 7th International Vacuum Congress and 3rd International Conference on Solid Surfaces (1977) 2447
345. Marbrow, R. A., Lambert, R. M.: Surface Sci. *61* (1976) 317
346. Gafner, G., Feder, R.: Surface Sci. *57* (1976) 37
347. Nieuwenhuys, B. E., Somorjai, G. A.: Surface Sci. *72* (1978) 8
348. Taylor, J. L., Weinberg, W. H.: Journal of Vacuum Science Technology *15* (1978) 590
349. Bas, E. B., Hafner, P., Klauser, S.: Proc. 7th International Vacuum Congress and 3rd International Conference on Solid Surfaces (1977) 881
350. Miura, T., Tuzi, Y.: Proc. 2nd International Conference on Solid Surfaces (1974) 85
351. Peralta, L., Berthier, Y., Oudar, J.: Surface Sci. *55* (1976) 199
352. Andersson, S., Pendry, J. B., Echenique, P. M.: Surface Sci. *65* (1977) 539
353. Küppers, J.: Surface Sci. *36* (1973) 53
354. Mitchell, D. F., Sewell, P. B.: Proc. 7th International Vacuum Congress and 3rd International Conference on Solid Surfaces (1977) 963
355. Mitchell, D. F., Sewell, P. B., Cohen, M.: Surface Sci. *69* (1977) 310
356. Madden, H. H., Küppers, J., Ertl, G.: J. Chem. Phys. *58* (1973) 3401
357. Madden, H. H., Ertl, G.: Surface Sci. *35* (1973) 211
358. Madden, H. H., Küppers, J., Ertl, G.: Journal of Vacuum Science Technology *11* (1974) 190
359. Taylor, T. N., Estrup, P. J.: Journal of Vacuum Science Technology *10* (1973) 26
360. Taylor, T. N., Estrup, P. J.: Journal of Vacuum Science Technology *11* (1974) 244
361. Price, G. L., Sexton, B. A., Baker, B. G.: Surface Science *60* (1976) 506
362. Wilf, M., Dawson, P. T.: Surface Sci. *65* (1977) 399
363. Ducros, R., Merrill, R. P.: Surface Sci. *55* (1976) 227
364. Lambert. R. M., Comrie, C. M.: Surface Sci. *46* (1974) 61
365. Reed, P. D., Lambert, R. M.: Surface Sci. *57* (1976) 485
366. Lambert, R. M.: Surface Sci. *49* (1975) 325
367. Berthier, Y., Oudar, J., Huber, M.: Surface Sci. *65* (1977) 361

368. Bonzel, H. P., Ku, R.: J. Chem. Phys. *58* (1973) 4617
369. Marbrow, R. A., Lambert, R. M.: Surface Sci. *67* (1977) 489
370. Orent, T. W., Hansen, R. S.: Surface Sci. *67* (1977) 325
371. Ku, R., Gjostein, N. A., Bonzel, H. P.: Surface Sci. *64* (1977) 465
372. Reed, P. D., Comrie, C. M., Lambert, R. M.; Surface Sci. *59* (1976) 33
373. Reed, P. D., Comrie, C. M., Lambert, R. M.: Surface Sci. *72* (1978) 423
374. Reed, P. D., Comrie, C. M., Lambert, R. M.: Surface Sci. 64 (1977) 603
375. Sakurai, T., Hagstrum, H. D.: Journal of Vacuum Science Technology *13* (1976) 807
376. Lo, W. J., Chung, Y. W., Somorjai, G. A.: Surface Sci. *71* (1978) 199
377. Van Hove, M. A., Tong, S. Y., Elconin, M. H.: Surface Sci. *64* (1977) 85
378. Wang, G. C., Lu, T. M., Lagally, M. G.: Proc. 7th International Vacuum Congress and 3rd International Conference on Solid Surfaces (1977) A2726
379. Baker, J. M., Eastman, D. E.: Journal of Vacuum Science Technology *10* (1973) 223
380. Buchholz, J. C., Lagally, M. G.: Journal of Vacuum Science Technology *11* (1974) 194
381. Buchholz, J. C., Lagally, M. G.: Physical Rev. Lett. *35* (1975) 442
382. Besocke, K., Berger, S.: Proc. 7th International Vacuum Congress and 3rd International Conference on Solid Surfaces (1977) 893
383. Madey, T. E., Yates, J. T.: Surface Sci. *63* (1977) 203
384. Engel, T., Niehus, H., Bauer, E.: Surface Sci. *52* (1975) 237
385. Buchholz, J. C., Wang, G. C., Lagally, M. G.: Surface Sci. *49* (1975) 508
386. Van Hove, M. A., Tong, S. Y.: Physical Review Letters *35* (1975) 1092
387. Bauer, E., Engel, T.: Surface Science *71* (1978) 695
388. Avery, N. R.: Surface Sci. *41* (1974) 533
389. Steinbruchel, Ch., Gomer, R.: Surface Sci. *67* (1977) 21
390. Steinbruchel, Ch., Gomer, R.: Journal of Vacuum Science Technology *14* (1977) 484
391. Avery, N. R.: Surface Sci. *43* (1974) 101
392. Göpel, W.: Surface Sci. *62* (1977) 165
393. Marbrow, R. A., Lambert, R. M.: Surface Sci. *71* (1978) 107
394. Papp, H., Pritchard, J.: Surface Sci. *53* (1975) 371
395. Kirby, R. E., McKee, C. S., Roberts, M. W.: Surface Sci. *55* (1976) 725
396. Maire, G., Bernhardt, P., Légaré, P., Lindauer, G.: Proc. 7th International Vacuum Congress and 3rd International Conference on Solid Surfaces (1977) 861
397. Schwaha, K., Bechtold, E.: Surface Sci. *65* (1977) 277
398. Netzer, F. P., Wille, R. A.: Journal of Catalysis *51* (1978) 18
399. Netzer, F. P., Wille, R. A.: Proc. 7th International Vacuum Congress and 3rd International Conference on Solid Surfaces (1977) 927
400. Christmann, K., Ertl, G.: Surface Sci. *60* (1976) 365
401. Gland, J.: Surface Sci. *71* (1978) 327
402. Castner, D. G., Somorjai, G. A.: Surface Sci., in press
403. Ertl, G., Plancher, M.: Surface Sci. *48* (1975) 364
404. Hopkins, B. J., Watts, G. D.: Surface Sci. *44* (1974) 237
405. Engel, T., von dem Hagen, T., Bauer, E.: Surface Sci. *62* (1977) 361
406. Gillet, E., Chiarena, J. C., Gillet, M.: Surface Sci. *67* (1977) 393
407. Bridge, M. E., Marbrow, R. A., Lambert, R. M.: Surface Sci. *57* (1976) 415
408. Buchholz, J. C., Somorjai, G. A.: Surface Sci. *66* (1977) 573
409. Atanasoska, L. L., Buchholz, J. C., Somorjai, G. A.: Surface Sci. *72* (1978) 189
410. Brodén, G., Rhodin, T., Capehart, W.: Surface Sci. *61* (1976) 143
411. Papageorgopoulos, C. A., Chen, J. M.: Surface Sci. *39* (1973) 283
412. Eastman, D. E., Denuth, J. E.: Proc. 2nd International Conference on Solid Surfaces (1974) 827
413. Demuth, J. E.: Surface Sci. *69* (1977) 365
414. Dalmai-Imelik, G., Bertolini, J. C., Massardier, J., Rousseau, J., Imelik, B.: Proc. 7th International Vacuum Congress and 3rd International Conference on Solid Surfaces (1977) 1179

415. Bertolini, J. C., Dalmai-Imelik, G., Rousseau, J.: Surface Sci. *67* (1977) 478
416. Casalone, C., Cattania, M. G., Simonetta, M., Tescari, M.: Surface Sci. *62* (1977) 321
417. Horn, K., Bradshaw, A. M., Jacobi, K.: Journal of Vacuum Science Technology *15* (1978) 575
418. Schouter, F. C., Kaleveld, E. W., Bootsma, G. A.: Surface Science *63* (1977) 460
419. McCarty, J., Madix, R. J.: Journal of Catalysis *38* (1975) 402
420. McCarty, J. G., Madix, R. J.: Journal of Catalysis *48* (1977) 422
421. Abbas, N. M., Madix, R. J.: Surface Science *62* (1977) 739
422. Maire, G., Anderson, J. R., Johnson, B. B.: Proc. Roy. Soc. (London) A 320 (1970) 227
423. Kesmodel, L. L., Baetzold, R. C., Somorjai, G. A.: Surface Sci. *66* (1977) 299
424. Stair, P. C., Somorjai, G. A.: J. Chem. Phys. *66* (1977) 573
425. Weinberg, W. H., Deans, H. A., Merrill, R. P.: Surface Sci. *41* (1974) 312
426. Lang, B.: Surface Sci. *53* (1975) 317
427. Firment, L. E., Somorjai, G. A.: J. Chem. Phys. *66* (1977) 2901
428. Stair, P. C., Somorjai, G. A.: J. Chem. Phys. *67* (1977) 4361
429. Gland, J. L., Somorjai, G. A.: Surface Sci. *38* (1973) 157
430. Gland, J. L., Somorjai, G. A.: Surface Sci. *41* (1974) 387
431. Fischer, T. E., Kelemen, S. R.: Surface Sci. *69* (1977) 485
432. Fischer, T. E., Kelemen, S. R., Bonzel, H. P.: Surface Sci. *64* (1977) 157
433. Netzer, F. P.: Surface Sci. *52* (1975) 709
434. Bridge, M. E., Lambert, R. M.: Journal of Catalysis *46* (1977) 143
435. Bridge, M. E., Lambert, R. M.: Surface Sci. *63* (1977) 315
436. Ducros, R., Housley, M, Alnot, M., Cassuto, A.: Surface Sci. *71* (1978) 433
437. Chung, Y. W., Siekhaus, W., Somorjai, G. A.: Surface Sci. *58* (1976) 341
438. Somorjai, G. A., Kesmodel, L. L.: MTP Int. Rev. Sci., Phys. Chem Ser. Two *7* (1975)
439. Wood, E. A.: J. Appl. Phys. *35* (1964) 1306
440. Ellis, W. P., Schwoebel, R. L.: Surface Sci. *11* (1968) 82
441. Blakely, D. W., Somorjai, G. A.: Surface Sci. *65* (1977) 419
442. Mönch, W.: Adv. in Solid State Physics *13* (1973) 241
443. Kesmodel, L. L., Dubois, L. H., Somorjai, G. A.: Chem. Phys. Lett. *56* (1978) 267
444. Somorjai, G. A., Kesmodel, L. L.: Trans. Am. Cryst. Assoc. *13* (1977) 67

unit stereographic triangle. As can be seen from that figure all the stepped surfaces which have low Miller index type steps lie on the 100, 110 and 111 zone lines. For surfaces which lie inside the unit stereographic triangle the steps themselves have steps and this type of surface is classified as a kinked surface. The only kinked surface for which surface structures have been reported is the Pt(10,8,7) or Pt(S) – (111) × (310) surface. The real space drawings and LEED patterns of the platinum (111), (755) and (10,8,7) surfaces are shown in Fig. 5.2.

In calculating the stepped surface designations that are listed in Table 5.5, it was assumed that the surfaces were stable in a monatomic step configuration, which is generally the case for the clean surfaces. This can readily be verified by LEED. In LEED patterns of stepped surfaces the step periodicity is superimposed on the terrace periodicity resulting in the splitting of the terrace diffraction spots into doublets or triplets at certain beam voltages. The direction of the splitting is perpendicular to the step edge and the magnitude of the splitting is inversely proportional to the terrace width, so the terrace width can be obtained by measuring the splitting observed in the LEED pattern. The step height can be determined from the formula

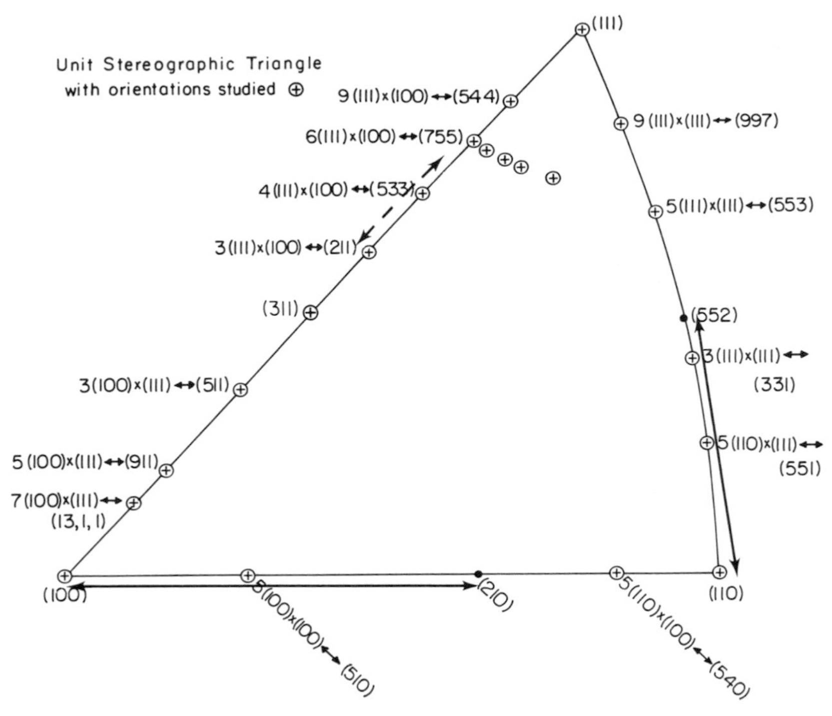

Fig. 5.1. The unit stereographic triangle for fcc showing the location and Miller indices of various faces

$$V_{00}\ (\text{singlet max}) = \frac{150}{4d^2\cos^2\phi}\ n^2$$

where V_{00} are the voltages for which a singlet intensity maximum of the (0,0) beam is observed, d is the step height, ϕ is the angle between surface normal and terrace normal, and n is an integer. By combining the terrace width and step height with the angle between the terrace and step planes the macroscopic surface plane can be determined.

The stability of stepped surfaces is an important consideration in LEED studies. Although these surfaces have higher surface free energies than the low index faces, most of the clean stepped surfaces are stable in a single step height configuration from room temperature to the melting point of the metals. When gases are adsorbed on these surfaces, however, their stability can noticeably change. Some surfaces reconstruct, forming multiple height steps and large terraces. Other high index surfaces form large low index facets while some retain the single step height configuration.

The surface structures observed for gas adsorption on stepped surfaces are listed in Table 5.6. In this table the stepped surfaces are denoted by either their Miller index label or stepped surface designation, depending on which system was used by the original author. By using Table 5.5 one may convert back and forth between these two systems. It is interesting to compare the surface structures formed on stepped surfaces to those

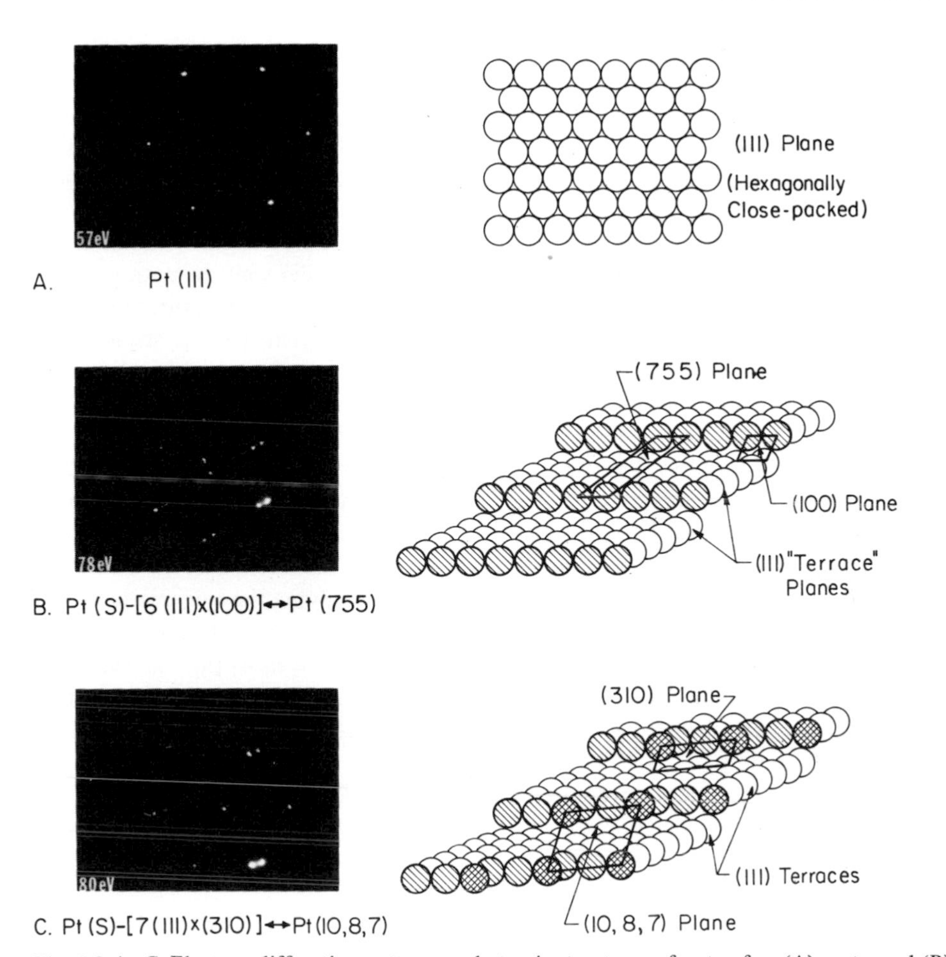

Fig. 5.2.A–C. Electron diffraction patterns and atomic structures of a step-free (A), a stepped (B) and a kinked (C) surface of platinum

formed on the low index faces given in Tables 5.2–5.4. For stepped surfaces with fairly large terrace widths (~6 to 8 atoms or larger) the surface structure that forms on the terrace is generally the same as the one that forms on the low index face. The surface structures on the low index surfaces tend to be more well ordered than those on the stepped surfaces. An example of this is the existence of several one-dimensional structures on stepped surfaces. The one-dimensional structures cause streaks to occur in the LEED patterns and are denoted as n-(1d) structures in the tables, where n is the number of streaks between rows of the substrate diffraction spots. Also, the adsorption of gases may cause faceting of the substrate due to the high surface free energy of stepped surfaces.

4 Ordered Organic Monolayers

The adsorption characteristics of organic molecules on solid surfaces is important in several areas of surface science. The nature of the chemical bonds between the substrate and the adsorbate, the ordering and orientation of the adsorbed organic molecules play important roles in adhesion, lubrication and hydrocarbon catalysis. Several studies have been undertaken to determine the molecular structure, ordering and interaction of monolayers for different groups of organic compounds under well-characterized conditions on low Miller index metal crystal surfaces. However the structures of only two of the small organic molecules, acetylene (C_2H_2) and ethylene (C_2H_4), adsorbed on the (111) crystal face of platinum has been determined so far using a combination of surface crystallography using the diffraction beam intensities measured by LEED, high resolution electron energy loss spectroscopy (HREELS) and ultra-violet photoelectron spectroscopy (UPS). These were discussed in Sect. III.

Over 50 other organic monolayers have been studied by LEED and a combination of other techniques when adsorbed on single crystal surfaces. Although structure analysis has not been carried out for these systems, their ordering characteristics, the size and orientation of their unit cell, have been determined. By studying the systematic variation of their shape and bonding characteristics, correlations can be made between these properties and their interactions with the metal surfaces. Analysis of the changes of surface structure with the shape or size of their unit cell often permitted unambiguous determination of their location and orientation on the surface even in the absence of surface crystallography.

We shall review the surface structures of monolayers of various homologues of organic compounds, the paraffins, the phthalocyanines, a few aromatic systems and amino acids that have been determined during recent investigations.

4.1 Normal Paraffin Monolayers and Thin Crystals or Platinum and Silver

Straight chain saturated hydrocarbon molecules from propane (C_3H_8) to octane (C_8H_{18}) were deposited from the vapor phase on platinum and silver(111) crystal surfaces in the temperature range 100–200 K. The ordered monolayer was produced first and then, with decreasing temperature a thick crystalline film was condensed and the surface structures of these organic crystals were also studied by LEED[1)].

At the highest temperature, T_1, at which the organic molecule condenses, at a given vapor flux, a surface structure is formed that exhibits only one-dimensional order. As the temperature is lowered these monolayer structures become more ordered and form a two-dimensional ordered surface structure at T_2. Upon further lowering the temperature to T_3, the rate of condensation of the organic vapor on the surface becomes greater than the rate of evaporation. At this temperature or below, the growth of organic multilayers commences. In Fig. 3.1 the phase transition temperatures for the various adsorbed paraffins are plotted as a function of chain length for adsorption and growth on the Pt(111) crystal face. The transition temperatures fall on a smooth curve and they increase with increasing chain length. Similar results have been obtained for deposition on the Ag(111) surface.

The paraffins adsorb with their chain axis parallel to the platinum substrate. Thus their surface unit cell increases smoothly with increasing chain length as shown in Fig. 5.3. The n-butane molecules, unlike the larger molecules, form several monolayer surface structures as the experimental conditions are varied. It appears that the smaller the paraffin the more densely packed it is on the surface. Evidently, as the packing becomes too dense for n-butane in one surface structure it forms a different one.

The monolayer adsorption characteristics of the C_3–C_8 paraffins are very similar on the Ag(111) to that on the Pt(111). The monolayers are less strongly held on the silver surface as manifested by the lower temperatures necessary to produce ordered surface structures on silver.

Multilayers condensed upon the ordered monolayers maintained the same orientation and packing as found in the monolayers. Thus, the monolayer structure determines the growth orientation and the surface structure of the growing organic crystal. This phenomenon is called pseudomorphism and as a result the surface structures of the growing organic crystals do not correspond to planes in the reported bulk crystal structures. The exception appears to be n-octane on the Ag(111) surface that is deposited with the $(10\bar{1})$ orientation of its bulk crystal structure.

The saturated hydrocarbons are very susceptible to electron beam damage, both in the monolayer and multilayer forms. While aromatic hydrocarbons and other conjugated systems exhibit minimal or no beam damage effects during the time necessary to carry out the LEED experiments, the ordered structures of paraffins disappear after ~5 sec of electron beam exposure as a result of desorption or partial dissociation of the organic adsorbates.

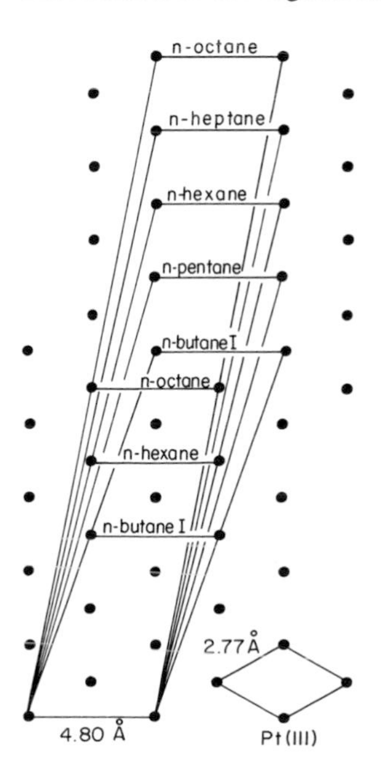

Fig. 5.3. Observed surface unit cells for n-paraffins on Pt(111)

4.2 Benzene, Cyclohexadiene, Cyclohexene, Cyclohexane and Naphthalene Monolayers on Platinum and Silver

Benzene adsorbs on the platinum(111) crystal face into a well ordered metastable $\begin{pmatrix}4 & -2\\ 0 & 4\end{pmatrix}$ structure at 300 K under ultrahigh vacuum conditions[2]. This initial structure transforms into a stable $\begin{pmatrix}4 & -2\\ 0 & 5\end{pmatrix}$ structure at a rate that is sensitive both to the sample temperature and to the flux of benzene vapor to the surface. I-V curves were taken from the diffraction beams of both surface structures and these indicate very little change in the carbon-platinum layer spacing during the structural transformation. However the work function changes with respect to the clean (111) platinum surface are –1.4 eV for the $\begin{pmatrix}4 & -2\\ 0 & 4\end{pmatrix}$ and –0.7 eV for the $\begin{pmatrix}4 & -2\\ 0 & 5\end{pmatrix}$ surface structures, a very large variation. Both the surface unit cell size and the calibrated Auger determination of the surface carbon content indicate that the adsorbate structure must have some of the benzene molecules inclined at an angle to the surface. In the absence of surface structure analysis the precise location of the benzene molecules with respect to each other or relative to the surface platinum atoms cannot be identified. However a complete set of I–V curves is available and should be a sufficient data base for structure analysis.

Benzene forms a rotationally disordered structure on the reconstructed (100) platinum surface. However, the work function changes with increasing surface coverage are similar to that of benzene on the (111) crystal face.

The adsorption of cyclohexadiene on the Pt(111) surface produces the same two surface structures that were found during the adsorption of benzene on this crystal face[2]. Thus, this molecule readily dehydrogenates on this platinum surface to benzene at 300 K.

Cyclohexene adsorbed[2] on the Pt(111) surface produces a $\begin{pmatrix}2 & 2\\ 4 & -2\end{pmatrix}$ surface structure at 300 K. The work function change upon adsorption is –1.7 eV. As the temperature is increased to 450 K a new $\begin{pmatrix}2 & 0\\ 0 & 2\end{pmatrix}$ surface structure appears.

Cyclohexane[3] forms a (9×9) surface structure on the Ag(111) crystal face and a $\begin{pmatrix}4 & -1\\ 1 & 5\end{pmatrix}$ surface structure on the Pt(111) crystal face at around 200 K. This latter surface structure corresponds to the (001) surface orientation of the monoclinic bulk crystal structure of the molecule. On heating the platinum crystal face to 450 K a $\begin{pmatrix}2 & 0\\ 0 & 2\end{pmatrix}$ surface structure forms that is identical to the surface structure formed by cyclohexene monolayers at the same temperature. It appears that cyclohexane dehydrogenates at elevated temperatures on platinum to form the same species or that of cyclohexene.

Naphthalene forms[2] a glassy, poorly ordered monolayer on Pt(111) at 300 K. However, upon heating to 450 K the monolayer orders to form a (6×6) surface structure. Adsorbed naphthalene forms a disordered layer also on the Ag(111) crystal face at 300 K. However, below 200 K an ordered structure appears with a unit cell $\begin{pmatrix}2.8 & 0.8\\ -2.0 & 3.8\end{pmatrix}$ and sometimes another, less stable monolayer structure is also detectable.

It is interesting that benzene and naphthalene form monolayer surface structures on the Pt(111) crystal face at 300 K and higher temperatures while monolayer surface structures form only at low temperatures (~200 K) on the Ag(111) crystal face[3]. While these aromatic molecules are held by strong chemical bonds to the platinum, their heats of adsorption must not be greater than the heats of sublimation

(10.7 and 17.3 kcal/mole for benzene and naphthalene, respectively) on the silver crystal plane. Thus adsorbate-adsorbate and adsorbate-substrate interactions are of the same magnitude for silver.

4.3 Other Organic Adsorbates on Platinum

The adsorption and ordering characteristics of a large group of organic compounds has been studied on the platinum(100) and (111) single crystal surfaces[2]. Low-energy electron diffraction has been used to determine surface structures. Work function change measurements have been made to determine the charge redistribution which occurs on adsorption. The molecules that have been studied are aniline, benzene, biphenyl, n-butyl-benzene, t-butylbenzene, cyanobenzene, cyclopentane, mesitylene, 2-methylnaphthalene, nitrobenzene, propylene, pyridine, toluene, and m-xylene. All of the molecules studied adsorb on both the Pt(111) and Pt(100) – (5 × 1) surfaces and are electron donors to the metal surface. The adsorbed layers are more ordered on the hexagonally symmetric Pt(111) surface than on the square symmetric Pt(100) surface. Unsaturated molecules generally adsorb on these crystal faces of platinum by forming π-bonds with the metal surfaces, indicated by work function change studies as well as the large heats of adsorption of these molecules when compared to the heats of adsorption of saturated hydrocarbons.

4.4 Phthalocyanine Monolayers and Films on Copper

Monolayer structures and epitaxial growth of vapor-deposited crystalline phthalocyanine films on single crystal copper substrates were studied using low energy electron diffraction[4]. Ordered monolayers of three different phthalocyanines, copper, iron, and metal-free, were seen on two different faces of copper, the (111) and (100). The monolayer structures formed were different on the two crystal faces and the several phthalocyanines yield nonidentical monolayer structures.

Ordered multilayer deposits were grown on both the Cu(111) and Cu(100) substrates. Electron beam damage to the phthalocyanine molecules was not observed. Space-charge effects due to electron bombardment were not apparent below an incident electron energy of 25 eV.

The surface structures observed for the multilayer deposits of the phthalocyanines on both substrate faces, Cu(111) and Cu(100), were not those of any plane in the bulk crystal structure of the phthalocyanines.

The monolayer surface structures observed for the various phthalocyanines on the two copper substrates are summarized in Table 5.7. In all cases, the size of the surface unit mesh is consistent with a surface structure unit cell containing a single planar phthalocyanine molecule oriented parallel to the substrate. The bonding to the copper substrate is largely through the phthalocyanine ligand rather than through the central metal atom, since the metal-free phthalocyanines are found, from thermal desorption experiments, to be bound as strongly to the Cu(100) and Cu(111) surfaces as the Cu- and Fe-phthalocyanines. While the central metal atom in the phthalocyanines has no effect on the surface structures formed on Cu(100), it does play a major role in determining the surface structure on Cu(111). Not only are the monolayer structures different for the three phthalocyanines, but the epitaxy of the multilayer deposit indicates a fundamental difference in the interaction of the metal and metal-free phthalocyanine with the Cu(111) surface. The metal-free phthalocyanine film grows, in the multilayer deposits, as a number of individual domains or crystallites, each yielding its own diffraction beams including its own specular reflection, since the surface planes are not parallel to one another. The metal-free phthalocyanine film exhibits sixfold symmetry in crystallite orientation. The Cu(111) surface, although sixfold symmetric in the atomic positions in the top layer, is only threefold symmetric when the positions of the second and third layer atoms are included (. . . ABC . . . stacking of an fcc crystal). Thus the metal-free phthalocyanine interacts with the substrate surface either through nonlocalized interactions such as van der Waals forces or bonds with only electrons in the copper which exhibit sixfold symmetry, i.e., the metallic s electrons of the top copper layer. Copper- and iron-phthalocyanines exhibit threefold symmetry in crystallite growth. Thus the addition of a metal atom in the phthalocyanine reduces the apparent symmetry of the substrate. The central metal atom is thus involved in bonding to the second layer copper atoms or to threefold symmetric electron orbitals, for example, *d* orbitals, of the surface copper atoms.

Deposition of Cu-phthalocyanine on a Pt(111) surface resulted in only poorly ordered monolayer structures and no ordering of multilayer structures. This demonstrates the importance of the details of the adsorbate-substrate interaction even for very large adsorbates which overlap tens of surface atoms. The absence of an ordered multilayer structure on this substrate indicates the role of an initially ordered monolayer in controlling epitaxial growth.

Ordered multilayer deposits of phthalocyanine molecules could be observed by low-energy electron diffraciton with no apparent electron beam induced chemical effects. This appears consistent with the general trend for molecules with highly conjugated electron systems to be more stable under electron bombardment than other organic molecules.

The surface structures observed for the multilayer phthalocyanine films are summarized in Table 5.7. These structures do not correspond to planes of either of the previously reported crystal structures of vapor deposited phthalocyanine films since the unit mesh constants reported in this work are considerably smaller than those previously reported. The unit cell dimensions correspond much more closely to a unit cell containing one molecule rather than, for example, the four molecules per cell reported for α-phthalocyanine.

4.5 Amino Acid Monolayers and Films on Copper

Monolayer structures and ordered multilayer films of several amino acids on single-crystal substrates were studied using low-energy electron diffraction[5)]. At monolayer coverage, ordered layers of glycine, alanine, D- and L-tryptophan were observed on both Cu(100) and Cu(111). For both glycine and alanine on Cu(111) the unit cell size suggests several molecules per unit cell, considering the dimensions of the nearly close-packed ac plane in bulk glycine. The alanine unit cell on Cu(100) is consistent with a single molecule per unit cell. The unit cell for glycine on Cu(100) requires at least two molecules per unit cell.

The monolayer structures for tryptophan are consistent with one and two molecules per unit cell, respectively, for Cu(100) and Cu(111). The structures observed for D- and L-tryptophan are related by mirror inversion which is consistent with the symmetry relationship between the two molecules. A mixture of the optical isomers, DL-tryptophan does not form an ordered monolayer, thus there is no segregation or cooperative interaction between the different isomers.

In addition to forming ordered monolayer structures, ordered multilayer films of several hundred Angstrom thickness were also grown for tryptophan. Ordered multilayers could be grown for DL-tryptophan even though the DL-tryptophan monolayer was disordered.

Electron beam damage effects followed the general rule that molecular groups in intimate contact with the metal substrate and aromatic groups appear relatively stable. Thus in the monolayer, alanine, with a methyl group likely sticking out from the surface, was the only molecule found to be unstable. In multilayer films, only tryptophan with the aromatic indole group to stabilize the molecule was found to yield multilayers stable under electron beam irradiation.

References

1a. Firment, L. E.: Ph.D. Thesis. University of California, Berkeley, (1976)
b. Firment, L. E., Somorjai, G. A.: J. Chem. Phys. *66*, 2901 (1977)
2a. Gland, J. L., Somorjai, G. A.: Adv. in Colloid and Interface Sci. *5*, 203 (1976)
b. Stair, P. C., Somorjai, G. A.: J. Chem. Phys. *67*, 4361 (1977)
3. Firment, L. E., Somorjai, G. A.: J. Chem. Phys. *69*, 3940 (1978)
4. Buchholz, J. C., Somorjai, G. A.: J. Chem. Phys. *66*, 573 (1977)
5. Atanasoska, L. L., Buchholz, J. C., Somorjai, G. A.: Surf. Sci. *72*, 189 (1978)

VI Surface Crystallography of Ordered Monolayers of Atoms

1 Introduction

By the use of mainly LEED and lately ion scattering techniques the location of many atomic adsorbates, their bond distances and bond angles from their nearest neighbor atoms have been determined. The substrates utilized in these investigations were low Miller Index surfaces of fcc, hcp and bcc metals in most cases, and low Miller Index surfaces of semiconductors that crystallize in the diamond, zincblende and wurtzite structures in some cases that could be cleaned and ordered with good reproducibility.

Since the substrate on which adsorbates are deposited greatly influences the behavior of those adsorbates, it is important to first examine the substrates themselves. We must distinguish between the clean surface and the same when covered with adsorbates, because adsorbates are capable of modifying the geometric (and electronic) structure of the substrate. To enable a convenient comparison, Table 6.1 combines the structures known to us for both clean and adsorbate-covered surfaces, as far as they have been determined with a reasonable degree of precision and reliability by the various surface crystallographic techniques mentioned in Section IV (co-adsorption and molecular adsorption are treated in the next Section).

2 The Effect of the Adsorbate on the Substrate Surface Structure

We shall here distinguish between surfaces that, in the clean state, have reconstructed or have unreconstructed structures. In the case of reconstructed structures, the surface atoms have moved sufficiently far away from their ideal bulk positions to either generate superlattices (i.e., larger two-dimensional structural unit cells) or, if no superlattice is present, at least substantially modified bond lengths or bond angles.

The general rule governing the small atomic displacements on clean unreconstructed surfaces seems to be that bond lengths increase slightly with the number of nearest neighbors (called the coordination number), in accordance with long-established knowledge[119]. Thus a surface atom, having lost some nearest neighbors compared to the bulk, tends to have a reduced bond length to its neighbors. Since the lattice constant parallel to the surface in the top layer is usually forced upon the top layer by the substrate, only the bond lengths to the second-layer atoms should in general decrease. This effect is small (at most about 4% contraction of bond lengths) and is sometimes

Table 6.1. Results of surface crystallography for clean surfaces and atomic adsorption, classified by structural type of substrate surface (alphabetical order is used within classes, considering only the letters of the chemical species). First the cases with metal substrates are listed, in order of decreasing close-packedness of the metal surfaces; these are then followed by the other materials. For metal substrates, the second column indicates relative bond length changes between first and second layers of the clean surface or the substrate and the corresponding layer spacing changes [between square brackets], referred to the bulk values: expansions and contractions have positive and negative signs, resp. (values close to 0% are mostly quoted as 0%, mainly when no variation away from 0% was tried in LEED calculations). In the adsorption site description, the stacking sequence for fcc(111) and hcp(0001) is indicated by the familiar ABCABC . . . or ABABAB . . . notation, lower-case letters being used for adsorbates; for other substrates the adsorption site is often characterized by the coordination number ("n-fold" meaning n nearest neighbors). Adsorption bond lengths are compared with bond lengths known from other sources for molecules and bulk compounds (to be found in the standard structure and crystallographic tables). Analytical methods are abbreviated as follows (see also Sect. II.1): KLEED, DDLEED, QDLEED, DLEED and SPLEED for Kinematical, Double-Diffraction (an approximation), Quasi-Dynamical, Dynamical and Spin-Polarized LEED; GVB, EH and Xα for Generalized-Valence-Bond, Extended Hückel and Xα cluster calculations

Surface type and surface	Topmost substrate bond length [and layer spacing] relaxation	Adsorption site	Adsorption layer spacing (Å)	Adsorption bond length (Å)	Equivalent bond lengths for non-surfaces	Method	References	Comments
fcc(111)								
Al(111)	−1 to +1.5% [−3 to +5%]					DLEED	1)	
	−1% [−3%]					KLEED & Fourier transf.	1e)	
Ag(111)	0% [0%]					DLEED	2)	
						KLEED & averaging	3)	
Ag(111) + p(1 × 1)Au	0% [0%]	cABC . . .	2.36	2.88	2.88	DLEED	2b)	
Ag(111) +	0% [0%]	cABC . . .	2.25	2.80	2.54–2.85	DLEED	4)	
$(\sqrt{3} \times \sqrt{3})R30^\circ$ I	0% [0%]	cABC . . .			2.54–2.85	Xα	5)	good results using DLEED geometry; cABC . . . better than bABC . . .
		b or cABC . . .	2.34	2.87 ± 0.02	2.54–2.85	SEXAFS	6)	
Ag(111) + incommensurate Xe	0% [0%]	variable	3.5	variable	3.53	KLEED & Fourier transf.	7)	
	0% [0%]	variable	3.55	variable	3.53	DLEED	8)	
Au(111)	0% [0%]					DLEED	2b)	reconstructions often observed
Co(111)	0% [0%]					DLEED	9)	high-temperature phase

Table 6.1 (continued)

Surface type and surface	Topmost substrate bond length [and layer spacing] relaxation	Adsorption site	Adsorption layer spacing (Å)	Adsorption bond length (Å)	Equivalent bond lengths for non-surfaces	Method	References	Comments
Cu(111)	−1.3 to 0% [−4 to 0%]					DLEED	10)	
Ir(111)	−0.8 ± 1.6% [−2.5 ± 5%]					DLEED & KLEED & Fourier transf.	11)	
Ni(111)	0% [0%]					DLEED	12)	
						KLEED & averaging	13)	
Ni(111) + p(2 × 2)2H	0% [0%]	b and cABC . . .	1.15 ± 0.10	1.84 ± 0.06	1.47 – 1.87	DLEED	14)	graphite overlayer geometry
		a, b, and cABC . . .	1.42 – 2.02	2.02 – 2.48	1.47 – 1.87	model calc.	15)	
Ni(111) + p(2 × 2)O	0% [0%]	b or cABC . . .	1.20 ± 0.10	1.88 ± 0.06	1.84 – 2.18	DLEED	16)	
Ni(111) + p(2 × 2)S	0% [0%]	cABC . . .	1.40 ± 0.10	2.02 ± 0.06	2.10 – 2.23	DLEED	17)	
	0% [0%]	cABC . . .	1.57	2.13	2.10 – 2.23	EH	18)	
Pt(111)	0% [0%]					DLEED & SPLEED	19)	
						MEIS & HEIS	20)	
Rh(111)	−0.3 ± 0.6% [−1 ± 2%]					DLEED	21)	
hcp(0001)								
Be(0001)	0% [0%]					DLEED	22)	
Cd(0001)	0% [0%]					DLEED	23)	
Co(0001)	0% [0%]					DLEED	9)	low-temperature phase
Ti(0001)	−0.5% [−2%]					DLEED	24)	
Ti(0001) + p(1 × 1)Cd	0% [0%]	cABA . . .	2.57 ± 0.05	3.08 ± 0.03	3.01	DLEED	23, 25)	
Ti(0001) + p(1 × 1)N	+1% [+5%]	AcBAB . . .	1.22 ± 0.05	2.095 ± 0.03	2.12	DLEED	26)	underlayer in octahedral holes
Zn(0001)	−0.5% [−2%]					KLEED & averaging	27)	
bcc(110)								
Fe(110)	0% [0%]					SPLEED	28)	

Na(110)	0% [0%]					SPLEED	29)	
W(110)	0% [0%]					KLEED & averaging	30)	
	0% [0%]					DLEED	31)	
W(110) + p(2 × 1)0	0% [0%]	3-fold	1.25 ± 0.10	2.08 ± 0.07	1.75 – 2.12	DLEED	32)	central bridge site not excluded
fcc(100)								
Al(100)	0% [0%]					DLEED	1b, 1c, 33)	
	0% [0%]					MEED	34)	
	0% [0%]					KLEED & Fourier transf.	1e)	
Al(100) + c(2 × 2) Na	0% [0%]	4-fold	2.06 ± 0.11	2.88 ± 0.08	2.82 – 3.00	DLEED	35)	
Ag(100)	0% [0%]					DLEED	36)	
Ag(100)+ c(2 × 2)Cl	0% [0%]	4-fold	1.72 ± 0.10	2.67 ± 0.06	2.36 – 2.77	DLEED	37)	
Ag(100) + c(2 × 2)Se	0% [0%]	4-fold	1.91 ± 0.10	2.80 ± 0.07	2.46 – 2.86	DLEED	38)	
Au(100)	0% [0%]					SPLEED	39)	metastable surface
Co(100)	–1.5% [–4%]					DLEED	40)	
Co(100) + c(2 × 2)O	0% [0%]	4-fold	0.80	1.94	2.12	DLEED	41)	
Cu(100)	0% [0%]					DLEED	10a, 42, 43)	
	0% [0%]					KLEED & averaging	44)	
	0% [0%]					KLEED & Fourier transf.	1e)	
Cu(100) + c(2 × 2)N	0% [0%]	4-fold	1.45 ± 0.04	2.32 ± 0.03	1.993 – 2.11	KLEED & averaging	45)	layer spacing of 0.90 is 2nd choice
	0% [0%]	4-fold	0.90 ± 0.10	2.02 ± 0.05	1.993 – 2.11	DLEED	46)	poor agreement between theory and experiment
Cu(100) + p(2 × 2)Te	0% [0%]	4-fold	1.70 ± 0.15	2.48 ± 0.10	2.51 – 2.76	DLEED	47	
Ni(100)	0% [0%]					DLEED	12a, 12c, 48)	
	0% [0%]					KLEED & Fourier transf.		
Ni(100) + p(2 × 2)C	+4% [+8.5%]					DLEED	49)	C position unknown; expansion confirmed by HEIS
Ni(100) + c(2 × 2)Na	0% [0%]	4-fold	2.23 ± 0.11	2.84 ± 0.08	2.80 – 3.10	DLEED	50, 51)	
Ni(100) + c(2 × 2)O	0% [0%]	4-fold	0.90 ± 0.10	1.98 ± 0.05	1.84 – 2.18	DLEED	52)	
		4-fold	0.90 ± 0.20	1.98 ± 0.10	1.84 – 2.18	LEIS	53)	
						INS	54)	O embedded in top Ni layer

Table 6.1 (continued)

Surface type and surface	Topmost substrate bond length [and layer spacing] relaxation	Adsorption site	Adsorption layer spacing (Å)	Adsorption bond length Å	Equivalent bond lengths Å	Method	References	Comments
Ni(100) + p(2 × 2)O	0% [0%]	4-fold	0.90 ± 0.10	1.98 ± 0.05	1.84 – 2.18	DLEED	55)	this geometry good in ARUPS theory vs. experiment
	0% [0%]	4-fold	0.96	2.01	1.84 – 2.18	GVB	56)	
	0% [0%]	bridge	1.20 ±0.10	2.13 ± 0.05	1.84 – 2.18	EH	18)	4-fold site expected with larger cluster
	0% [0%]	4-fold	0.75	1.91	1.84 – 2.18	Xα	57)	
Ni(100) + c(2 × 2)S	0% [0%]	4-fold	1.30 ± 0.10	2.19 ± 0.06	2.10 – 2.23	DLEED	52, 58)	this geometry good in ARUPS theory vs. experiment
		4-fold	1.30	2.19	2.10 – 2.23	LEIS	50)	
		4-fold				INS	60)	substrate hollow distorted (diamond shape)
Ni(100) + p(2 × 2)S	0% [0%]	4-fold	1.30 ± 0.10	2.19 ± 0.06	2.10 – 2.23	DLEED	55)	
	0% [0%]	4-fold	1.33	2.21	2.10 – 2.23	GVB	56)	
	0% [0%]	4-fold	1.31	2.20	2.10 – 2.23	EH	18)	
Ni(100) + c(2 × 2)Se	0% [0%]	4-fold	1.45 ± 0.10	2.28 ± 0.06	2.31 – 2.53	DLEED	52)	
	0% [0%]	4-fold	1.47	2.29	2.31 – 2.42	EH	18)	
Ni(100) + p(2 × 2)Se	0% [0%]	4-fold	1.55 ± 0.10	2.34 ± 0.07	2.31 – 2.53	DLEED	55)	
Ni(100) + c(2 × 2)Te	0% [0%]	4-fold	1.90 ± 0.10	2.58 ± 0.07	2.54 – 2.85	DLEED	52, 61)	
Ni(100) + p(2 × 2)Te	0% [0%]	4-fold	1.80 ± 0.10	2.52 ± 0.07	2.54 – 2.85	DLEED	55)	
Pt(100)	0% [0%]					SPLEED	62)	metastable surface
Rh(100)	0% [0%]					DLEED	63)	
Xe(100)	0% [0%]					KLEED	64)	
bcc(100)								
Fe(100)	–0.7 to –1.5% [–1.4 to –4%]					DLEED & SPLEED	65)	less contraction when less clean
Fe(100) + p(1 × 1)O	+3%[+7.5%]	4-fold	0.48	2.02 & 2.08	2.09 – 2.15	DLEED	66)	O closest to 2nd layer Fe atoms
		4-fold	0.48	2.02 & 2.08	2.09 – 2.15	EH	67)	
Fe(100) + (2 × 2)S	0% [0%]	4-fold	1.15 ± 0.05	2.33	1.99 – 2.44	DLEED	68)	

Mo(100)	−4% [−12%]					DLEED		[−15%] found by surface resonance analysis cf Ref. 69d)
Mo(100) + c(2 × 2)N	0% [0%]	4-fold	1.02 ± 0.10	2.45 ± 0.05	2.11 − 2.33	DLEED	70)	
Mo(100) + p(2 × 1)O	0% [0%]	4-fold	0.70	2.28 & 2.33	1.66 − 2.07	DLEED	71)	
Mo(100) + p(1 × 1)S:	0% [0%]	4-fold	1.16 ± 0.10	2.51 ± 0.05	2.53	DLEED	72)	
W(100)	−2 to −4% [−4.4 to −11%]					DLEED & SPLEED	31a, 31b, 73)	
	0 to −2.5% [0 to −6%]					HEIS	74)	
W(100)c(2 × 2)	[−6%]					DLEED	75)	low-temperature reconstruction: zig-zag rows of touching top-layer W atoms
W(100) + c(2 × 2)H	[−6%]					DLEED	75a, 76)	probably substrate identical to W(100)c(2 × 2)
W(100) + p(1 × 1)2H	0% [0%]					DLEED	76)	H probably in bridge sites
fcc(110)								
Al(100)	−3 to −4.5%					DLEED	1a-d, 33)	
	[−9 to −15%]					KLEED & Fourier transf.	77)	
						MEED	34)	
						model calc.	78)	
Ag(110)	−2 to −3%					DLEED	36b, 79)	
	[−6 to −10%]					KLEED & Fourier transf.	77)	
Ag(110) + p(2 × 1)O						DLEED	80)	probably long-bridge site
						LEIS	81)	probably long-bridge site
Cu(110)	−3 to −4% [−10 to −12%]					DLEED	82)	2nd layer spacing may be also contracted
Ir(110)(2 × 1)	−2% [−10%]					DLEED	83)	missing row reconstruction; 2nd layer atoms pressed sideways somewhat
Ir(110)(1 × 1)	−2.5% [−7.5%]					DLEED	84)	quarter-monolayer of randomly positioned O prevents reconstruction
Ir(110) + c(2 × 2)O	0% [0%]	short bridge	1.37 ± 0.05	1.93 ± 0.05		DLEED	85)	1/4-monolayer of randomly positioned O present

Table 6.1 (continued)

Surface type and surface	Topmost substrate bond length [and layer spacing] relaxation	Adsorption site	Adsorption layer spacing (Å)	Adsorption bond length (Å)	Equivalent bond lengths for non-surfaces	Method	References	Comments
Ni(110)	−1.5% [−5%]					DLEED	12a, 48)	
	−1.2% [−4%]					KLEED & Fourier transf.	77)	
	−1.2% [−4%]					MEIS	86)	
	+0.3% [+1%]					MEIS	86)	1/3 monolayer of randomly positioned O present
Ni(110) + p(2 × 1)H	−2.5% [−8%]					DLEED	87)	(2 × 1) structure probably due to substrate reconstruction (row pairing); H position unknown
Ni(110) + p(2 × 1)0	0% [0%]	short bridge	1.46 ± 0.05	1.92 ± 0.04	1.84 – 2.18	DLEED	88)	
		short bridge			1.84 – 2.18	LEIS	89, 90)	
Ni(110) + c(2 × 2)S	0% [0%]	center	0.93 ± 0.10	2.17 ± 0.10	2.10 – 2.23	DLEED	17a, 17b)	S closest to 2nd layer Ni atoms
	+6 ± 3% [+1.5 ± 0.75%]	center	0.87 ± 0.03	2.11 ± 0.03	2.10 – 2.23	MEIS	91)	
	0% [0%]	long bridge	1.04	2.04	2.10 – 2.23	GVB	92)	
Rh(110)	−0.9% [−2.7%]					DLEED	93)	
bcc(111)								
Fe(111)	−1.5% [−15%]					DLEED	94)	
fcc(311)								
Cu(311)	−1% [−5%]					DLEED	95)	

Surface type and surface	Surface structure, including bond length [and layer spacing] relaxations, relative to bulk	Method	References	Comments
Other Surfaces				
CoO(111)	O-terminated polar face of NaCl structure; top layer contraction −5% [−15%]	DLEED	96)	

GaAs(100)	Bulk structure with As termination, no relaxation	DLEED		As termination forced by Molecular Beam Epitaxy
GaAs(110)	Zincblende structure with top Ga and As atoms rotated into, resp. out of surface, (keeping about constant mutual bond length, rotated by projected angle of 27°); Ga and As back bonds contracted by –2.5% and –3.6%, resp.	DLEED	98)	
		model calc.	99)	agree qualitatively with LEED result
GaAs(110) + (1 × 1)As	Substrate has unrelaxed bulk structure; As bonded to surface Ga as in bulk (possible small bond angle change)	QDLEED	97a)	
MgO(100)	Unrelaxed bulk NaCl structure within [±5%]	DLEED	100)	
MoS_2(0001)	Layer compound cleaved between two 3-plane layers; top contraction by –1.6% [–4.7%], first Van der Waals spacing contracted [–3%]	DLEED	101, 32b)	
Na_2O(111)	Fluorite structure terminated between two Na layers; no relaxation	DLEED	102)	
$NbSe_2$(0001)	as MoS_2(0001), but top contraction by –0.2% [–0.6%], first Van der Waals spacing contracted [–1.4%]	DLEED	101, 32b)	
NiO(100)	Unrelaxed bulk NaCl structure within [±5%]	DLEED	103)	
Si(111)"(1 × 1)"	Bulk structure with –2% [–15%] top contraction.	DLEED	104)	impurity-stabilized
	Bulk structure with +0.8% [+5%] and +1% [+1%] expansions in two topmost layer spacings, resp.	DLEED	105a)	annealed at conversion temperature
	Bulk structure with –4% [–30%] top contraction	MINDO cluster	106a)	
	Bulk structure with –1% [–6%] top contraction.	model calc.	105b)	
Si(111)p(2 × 1)	Top layer contracted –1% [–8%], buckled ±3% [±22%]; 2nd layer spacing contraction –10% [–10%]	DDLEED	105a)	
	Qualitatively as above	MINDO cluster	106)	
	as DDLEED, but buckled ±10% [±50%], no 2nd layer spacing contraction	model calc.	106b)	
Si(111)(7 × 7)	Top double layer may tend toward one planar (graphite) layer, buckled and coincident with substrate with period (7 × 7)	KLEED	107)	
Si(100)p(2 × 1)	Top atom pairing (Schlier-Farnsworth model) with elastic relaxations down several layers	KLEED	108)	
		QDLEED	109)	
		DLEED	110)	
		model calc.	111)	
Si(100) + p(2 × 1)H	Probably: Substrate as Si(100)p (2 × 1)	model calc.	111)	
Si(100) + p(1 × 1)2H	Unrelaxed unreconstructed substrate	DLEED	112)	
TiS_2(0001)	As MoS_2, but top contraction by –1.7% [–5%], first Van der Waals spacing contracted [–5%]	DLEED	113)	
$TiSe_2$(0001)	As MoS_2, but top expansion by +1.7% [+5%]. first Van der Waals spacing contracted [–5%]	DLEED	113)	

Table 6.1 (continued)

Surface type and surface	Surface structure, including bond length [and layer spacing] relaxations, relative to bulk	Method	References	Comments
ZnO(0001)	Unreconstructed Zn-terminated wurtzite structure with top contraction by −3% [−25%]	DLEED	114)	
ZnO(1010)	Unreconstructed wurtzite structure, top Zn and O pulled into surface somewhat	DLEED	115-117)	
ZnSe(110)	Zincblende structure reconstructed as GaAs(110)	DLEED	118)	
	Qualitatively as above	model calc.	99d, 106b	

presumably drowned in the experimental and theoretical uncertainty (1 to 2% of these bond lengths) of the LEED analyses that have produced most of these results. In general, however, the effect is clearly discernible. A closer look at Table 6.1 shows that on clean close-packed faces – such as the fcc(111), hcp(0001), bcc(110) and fcc(100) faces of metals – almost no contraction is usually detected (rare cases of very small expansions are nevertheless reported). On less close-packed faces – such as the bcc(100), fcc(110), bcc(111) and fcc(311) faces of metals – small contractions are systematically detected in LEED analyses (cf. Fig. 6.1). Such results find independent confirmation in ion scattering experiments and theoretical calculations (cf. references in Table 6.1). They are also in qualitative agreement with very small (~1%) bond length contractions observed, e.g., in electron diffraction studies of 12 to 92 Å radius metal clusters[120].

The physical or chemical origin of these contractions can be explained in different terms. First-principles descriptions are too involved for inclusion here (see Ref. [32a] of Sect. IV). Instead we indicate some phenomenological descriptions. Firstly one can imagine the electron cloud to attempt to smooth its surface (as if there were a surface tension), thereby producing electrostatic forces that draw the surface atoms towards the substrate. This effect should be the stronger the less close-packed the surface is[78]. Secondly, with fewer neighbors the two-body repulsion energy is smaller, allowing greater atomic orbital overlap and therefore more favorable bonding at shorter bond

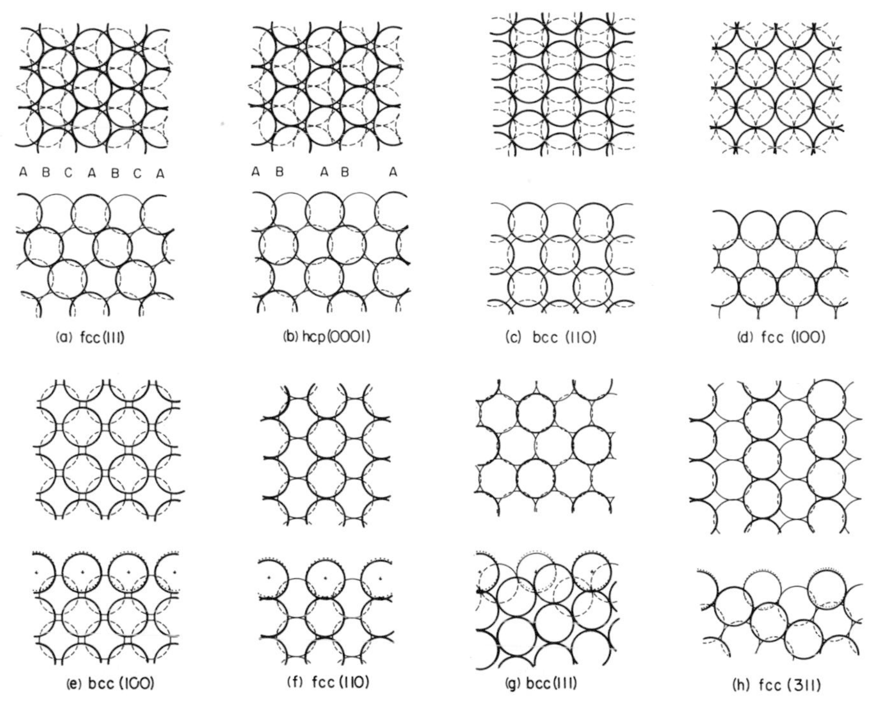

Fig. 6.1. Atomic arrangement in various clean metal surfaces. In each panel (*a–h*) the top and bottom sketches give top and side views, respectively. Thin-lined atoms are behind the plane of thick-lined atoms. Dotted lines represent atoms in unrelaxed (ideal bulk) positions; relaxations are shown by arrows

lengths. Thirdly one may say that for surface atoms the bonding electrons are partly shifted from the cut bonds to the remaining non-cut bonds, thereby increasing the charge content of the latter and so reducing the bond length. On ionic crystal surfaces the asymmetries in the ionic electrostatic forces at surfaces may explain the contrast between similar bond length contractions observed on CoO(111) and the lack of observable contractions on MgO(100) and NiO(100)[121].

The above descriptions of the origin of bond length contractions at surfaces are consistent with the observations made when adsorbates are deposited on these surfaces: the shortened bond lengths are systematically lengthened again (sometimes to more than their bulk values) by the presence of adsorbates, as is visible in Table 6.1 (cf. Fig. 6.2).

Surprisingly only a half monolayer of adsorbates is often sufficient to restore the bulk bond length between the substrate atoms. This behavior is observed both by LEED and by ion scattering experiments. With Fe(100) + p(1 × 1)O the underlying metal bond lengths are expanded to beyond their bulk value and in that process the FeO bulk oxide geometry is approached, presumably exhibiting a first stage of the

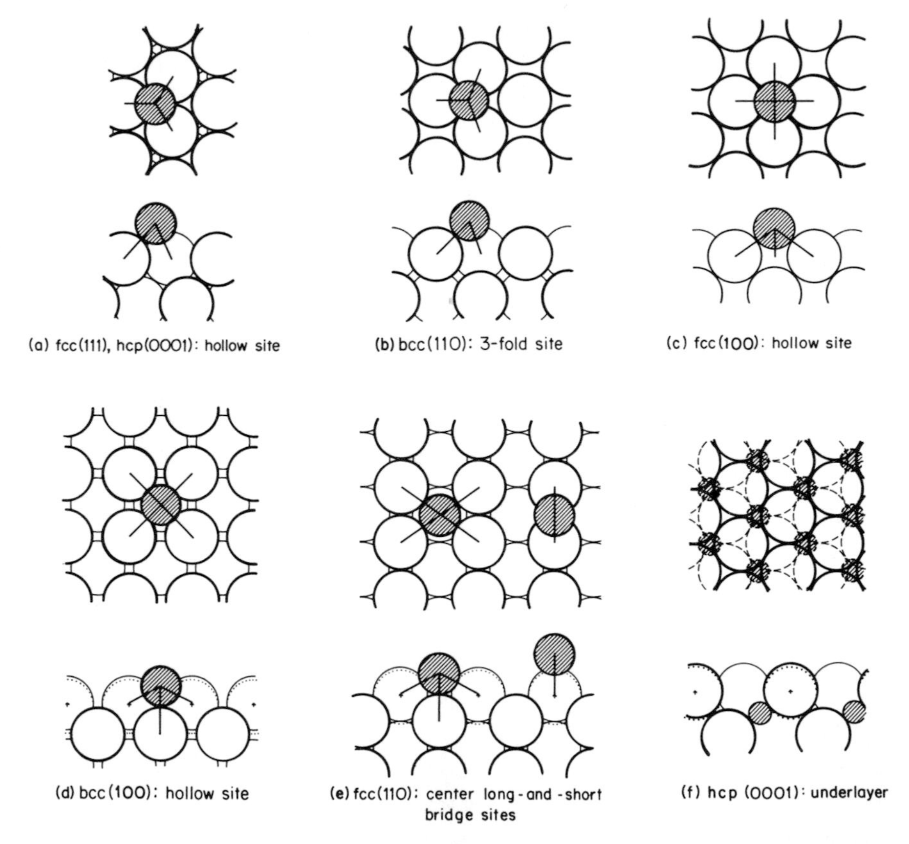

Fig. 6.2. Top and side views (in top and bottom sketches of each panel) of adsorption geometries on various metal surfaces. Adsorbates are drawn shaded. Dotted lines represent clean-surface (relaxed) atomic positions; arrows show atomic displacements due to adsorption

oxidation process at a surface. Ion scattering experiments indicate a similar behavior (not searched for in an earlier LEED analysis) for Ni(110) + c(2 × 2)S. In this connection it is interesting to note the case of Ti(0001) + (1 × 1)N, where a surface slab of three layers essentially identical to the bulk compound TiN is formed by a slight expansion of the topmost Ti-Ti layer spacing and intercalation of nitrogen (cf. Fig. 6.2f). Not properly understood yet is the case of Ni(100) + p(2 × 2)C in which an expansion of the topmost Ni-Ni layer spacing also occurs and possibly some kind of nickel carbide is formed. With O on Al(100) (not studied with LEED) there is considerable evidence for adatom penetration, i.e., metal oxide formation[122)]. Oxides can of course be formed on many surfaces, but details of geometry and behavior remain to be elucidated.

A variety of different reconstruction geometries are thought to occur on surfaces. On fcc metals large superlattices, e.g., (5 × 1) or (5 × 20), are observed on Pt(100),[123)] Ir(100)[124a)] and Au(100)[125)]. Some indications point to a hexagonally close-packed restructuring of the topmost atomic layer, but other structures are possible. A weakly bound adsorbate layer, such as that formed during the physisorption of Xe on Ir(100)[124b)], appears to not affect the basic geometry of the reconstructed substrate. However these reconstructions are usually destroyed in favor of the unreconstructed geometry as a result of chemisorption with its stronger substrate-overlayer bonding. This can even happen with rather small coverage, such as a few percent of a monolayer. In essence the adstoms simulate the "missing half" of the substrate. These "impurity-stabilized" unreconstructed surfaces [e.g., Pt(100) and Au(100)] have the structure known for the other stable clean unreconstructed metal surfaces.

The (110) face of these materials (Pt, Ir and Au) often exhibits a (2 × 1) reconstruction or more generally (n × 1) reconstructions [such as on Ir(110) with n = 2, 3 or 4][126)] with sometimes a statistical distribution of the values of the integer n (as on Au(110), where n = 2 dominates)[127)]. Several models for these reconstructions have been suggested, but the "missing row" model seems to be the most promising in studies of Ir(110) (2 × 1) (cf. Fig. 6.3) and Au(110) (n × 1)[127)] with random n: in this model small facets of the hexagonally close-packed (111) face are built [note the analogy with the fcc(100) reconstructions], which is consistent with the knowledge that the (111) face of fcc metals is energetically the most favorable one. These reconstructions also seem to be destroyed by adsorbates in favor of the unreconstructed structure. This has been established in particular with the Ir(110) surface. There the (2 × 1) reconstruction disappears as a result of adsorption of a disordered quarter monolayer of oxygen; in fact the resulting substrate shows contracted bond lengths just as with the unreconstructed clean fcc(110) surfaces. Adsorption of an additional half monolayer of oxygen, which orders in a c(2 × 2) arrangement, then removes that bond length contraction. This two-step process supports the notion that the effect of adsorbates on the substrate grows with the coverage. But apparently the effect is sometimes strong (a small fraction of a monolayer can destroy a reconstruction) and sometimes weak (a full film of xenon on Ir(100) seems not to destroy the reconstruction). A special behavior is found for hydrogen on Ni(110), which produces a (2 × 1) superlattice believed to be due to a pairwise attraction and approach of adjoining rows of surface nickel atoms (the row-pairing model).

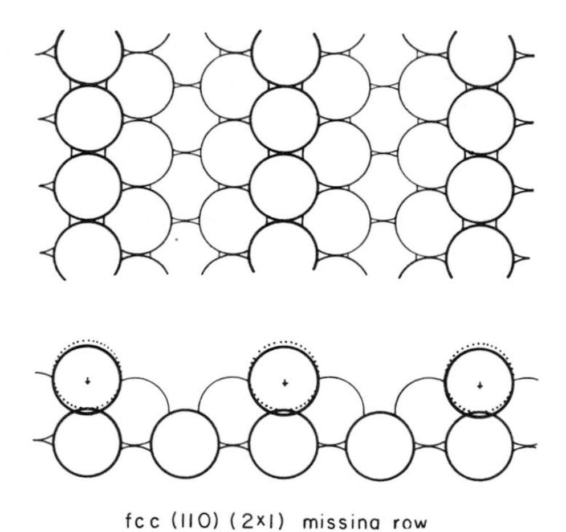

Fig. 6.3. Atomic arrangement in the missing-row model of the Ir(110)(2 × 1) surface. Conventions as in Fig. 6.1

Although the (111) face of fcc metals is of the lowest surface free energy – a fact which may explain the reconstructions of the (100) and (110) faces –, the (111) face itself may also reconstruct: Au(111) is normally reconstructed with a structure that may nevertheless still involve the hexagonally close-packed layer geometry (since extra sites of hexagonally arranged spots appear in LEED), but with a lattice constant different from that of the bulk[128].

On bcc metals only one reconstruction has been thoroughly analyzed, namely that of W(100)c(2 × 2), which occurs at low temperatures, (cf. Fig. 6.4). The mechanism responsible for this could be a charge density wave[129] that induces a structural wave which can have a wavelength related to the lattice constant [as with W(100)c(2 × 2)] or not related to it [as with Mo(100)[130]].

Hydrogen chemisorption at less than full coverage appears to not change the structure of the W(100)c(2 × 2) surface noticeably[75c]. Interestingly, chemisorption of hydrogen at room temperature on an unreconstructed W(100) surface seems to generate the same c(2 × 2) reconstruction obtained by simple cooling[75c, 76]. At full hydrogen coverage, the reconstruction disappears and W–W bond length contractions seem to disappear as well[76].

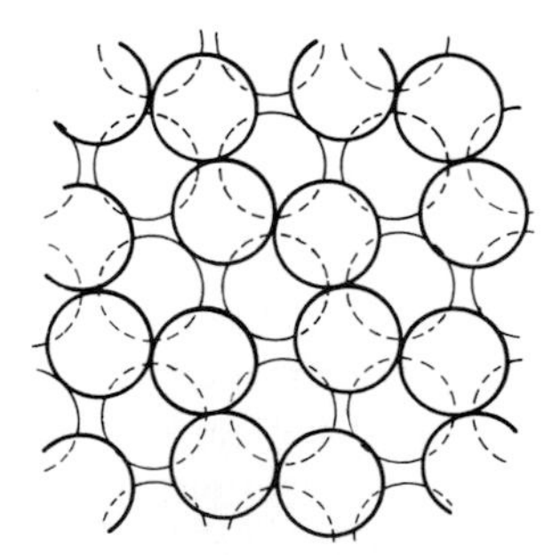

Fig. 6.4. Top view of the W(100)c(2 × 2) surface. Conventions as in Fig. 6.1

A type of reconstruction that one might expect to occur but that has not been observed is related to the relatively easy phase transition between hcp and fcc metals: this involves only the shifting of hexagonally close-packed layers of atoms, from the ... ABABAB ... to the ... ABCABC ... stacking arrangement. Such a shift could easily occur for the topmost atomic layer of hcp(0001) or fcc(111) surfaces. Interestingly, it does not seem to take place in reality on the five hcp(0001) and nine fcc(111) surfaces analyzed so far; this includes the case of Co on both sides of its hcp-fcc phase transition.

Reconstructions are particularly frequent on semiconductor surfaces. In three cases the structure and the underlying mechanism have most probably been identified. For GaAs(110) [and ZnSe(110), which behaves as GaAs(110), but whose properties have been less extensively studied], rehybridization of the orbitals around the surface Ga and As atoms occurs, producing new optimum bond angles (different from the tetrahedral angles of the bulk) that force substantial movements of the surface atoms (bond lengths remaining almost unchanged). In this case the atomic movements can be accommodated without enlarging the surface unit cell, so that no superlattice is generated, cf. Fig. 6.5. This type of reconstruction is also predicted by several model calculations. The adsorption of a monolayer of arsenic on this reconstructed surface restores the bulk lattice geometry in the topmost substrate layer, as a LEED analysis indicates (the adsorbed As atoms bond to the surface Ga atoms). Oxygen adsorption also has the same effect on GaAs(110). Oxygen (which appears to bond to the surface As atoms) approximately restores the bulk geometry of the surface. This is indicated both by UPS data[99b)] and by theoretical cluster calculations.[99c)]

For Si(100)p(2 × 1) a long search has produced a structure derived from the Schlier-Farnsworth model[131)], in which adjoining surface atoms (each with two unsatisfied "dangling bonds") simply bond together by bending over towards each other and pairing up dangling bonds (one dangling bond per surface atom remains unsatisfied). A substantial bond bending occurs and this distortion propagates elastically through the lattice down to a few layers' depth[108)], cf. Fig. 6.6. Adsorption of hydrogen up to a certain coverage onto this surface seems not to change the nature of this reconstruction. However a higher coverage of hydrogen destroys the reconstruction and restores the bulk geometry at the silicon substrate surface (apparently the surface Si–Si bonds have been broken and possibly replaced by bonds between the surface silicon atoms and the additional hydrogen atoms).

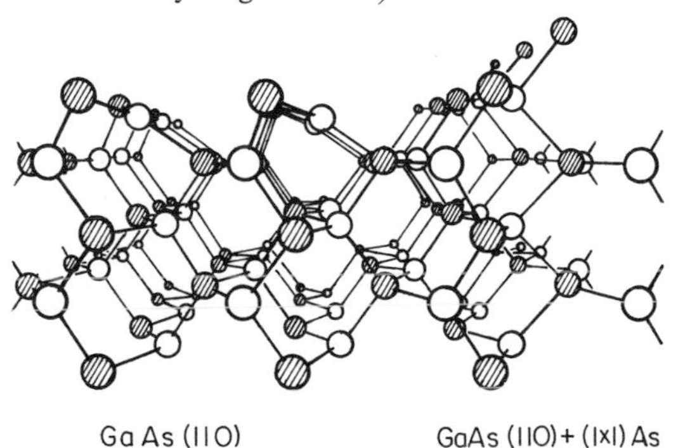

Fig. 6.5. Perspective view, looking along surface of clean reconstructed GaAs(110) at left and GaAs(110) + (1 × 1)As at right. Open and shaded circles represent Ga and As atoms, respectively

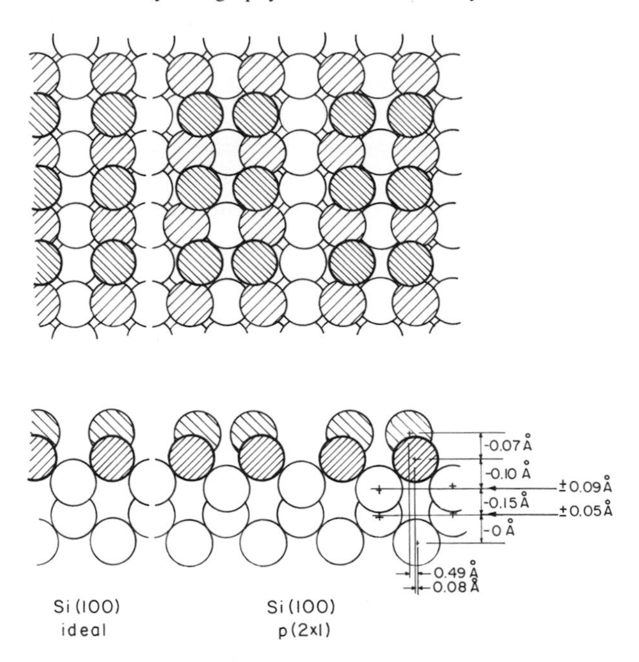

Fig. 6.6. Top and side views (in top and bottom sketches, respectively) of ideal bulk-like Si(100) at left and Si(100)p(2 × 1) in the modified Schlier-Farnsworth model at right. Layer spacing contractions and intra-layer atomic displacements relative to the bulk structure are given. Shading differentiates surface layers

The special influence of hydrogen on the substrate should be stressed. Whereas other adsorbates leave the substrate essentially unchanged or else remove a reconstruction, hydrogen at low coverages can induce a substrate reconstruction (at high coverages hydrogen behaves as other adsorbates).

3 The Adsorption Geometry of Atoms

With atomic adsorption on semiconductor surfaces one expects adsorbates to choose positions on the substrate that are relatively more predictable than with metal substrates: semiconductor surfaces often have relatively well-defined unsatisfied bonds (dangling bonds) ready to serve as adsorption sites, whereas such a simple argument does not seem to apply to metal surfaces. Such a behavior would be expected also from the geometry of bulk compounds, which tend to be relatively unique and predictable for semiconductors (e.g., the very common zincblende and wurtzite structures based on tetrahedral arrangements) but much more varied and complicated for compounds containing metal atoms (thus many different crystallographic phases of such compounds exist, e.g., with metal oxides). On the other hand the stronger tendency towards surface reconstructions for semiconductors compared to metals (possibly due to the lack of close-packing and ensuing freedom of movement in semiconductors, associated with a strong desire to satisfy stoichiometry) adds a different dimension to the structural possibilities. Thus the one established geometry for adsorption on a semiconductor, that for GaAs(110) + p(1 × 1)As, has the simple structure expected from the bulk geometry of GaAs, i.e., As bonds to the surface Ga atoms, cf. Fig. 6.5. With oxygen on GaAs(110), it appears from UPS[99b)] and cluster calculations[99f)] that the reconstruction is also removed. For hydrogen on Si(100), although the hydrogen positions have not been determined directly, the hypothesis of adsorption to the Si dangling bonds is consistent with the observations of the disappearance of the (2 × 1) reconstruction of the clean surface and with model calculations, cf. Fig. 6.6.

Turning to metal substrates, in most cases of atomic adsorption on metal surfaces where the adsorption geometry has been determined (cf. Table 6.1), only one adsorption site is involved, i.e., all adatoms have identical surroundings [the exceptions are Ni(111) + p(2 × 2)2H and Ag(111) + Xe, discussed below]. The adsorption can thus be conveniently characterized by the adsorption site and the metal-adsorbate bond lengths.

Adsorption sites (or "registries") on metals differ mainly in the number of nearest metal neighbors (the coordination number) and the two-dimensional symmetry. One might expect the adatom valence to influence the number of nearest metal neighbors and therefore the adsorption site. However there is little evidence for such a behavior, The divalent oxygen chooses the 2-fold coordinated short-bridge site on Ni(110) and Ir(110), while on Ag(110) the long-bridge site with an uncertain coordination number (depending on the unknown bond lengths) may be chosen. Oxygen on other surfaces and the divalent S, Se and Te, as well as all other adsorbates on various surfaces do not show this behavior. Instead, one finds the strong tendency for adatoms to occupy the sites with the largest available number of nearest metal neighbors, cf. Table 6.2 and Fig. 6.2. Even W(110) + p(2 × 1)O seems to involve the three-fold adsorption site rather than the higher-symmetry central two-fold site that one might predict from the oxygen valency or by using sites obtained by extending the substrate lattice out beyond the surface.

It is interesting to note that this tendency towards occupying the site with the largest coordination number during adsorption on metals holds [except with oxygen on fcc(110)] independently of the crystallographic face for a given metal, independently of the metal for a given crystallographic face and independently of the adsorbate for a given substrate.

It will be observed that adsorption sites with many nearest neighbors are usually also sites of high symmetry. Therefore one may say that adsorbate atoms appear to favor sites of high symmetry. There is only one exception to this preference: in W(110) + p(2 × 1)O the oxygen seems to choose a site that by itself (ignoring other adsorbed atoms) has one mirror plane instead of a site that has two orthogonal mirror planes; this may be related to the fact that the overlayer as a whole already has low symmetry (only a 2-fold axis of rotation). Even for oxygen on fcc(110) surfaces, a site with the highest possible symmetry is chosen: no other site has higher symmetry (although several have the same symmetry, two orthogonal mirror planes).

If one now also takes the second and deeper substrate layers into consideration, one may in particular wonder whether the adsorbate atoms choose an adsorption site consistent with a continuation of the substrate lattice. It appears from the available results that the bulk lattice is in fact usually continued into the overlayer, as if a substrate atom rather than a foreign atom had adsorbed, despite differing bonding characteristics. This bulk lattice continuation is satisfied by nearly all the cases listed in Table 6.1; O on fcc(110) again is an exception. Ni(111) + p(2 × 2)2H and Ti(0001) + p(1 × 1)Cd are also exceptions, belonging to an interesting class of surfaces. These are the hcp(0001) and fcc(111) surfaces, which we describe by the registry sequences ABABAB. . . and ABCABC. . . , respectively (surface at left). Using lower case letters for overlayers, the continuation of the bulk lattice into the overlayer would imply the sequences bABABAB . . . and cABCABC . . . , respectively. These sequences are indeed found for fcc Ag(111) + p(1 × 1)Au, fcc Ni(111) + p(2 × 2)S and fcc Ag(111) + ($\sqrt{3} \times \sqrt{3}$) R30° I. For fcc Ni(111) + p(2 × 2)O it could not be determined whether the sequence is cABCABC . . . or bABCABC . . . With fcc Ni(111) + p(2 × 2)2H and its two adatoms per unit cell, cf. Fig. 6.7, LEED predicts that both three-fold coordinated sites are used, i.e., both cABCABC . . . and bABCABC . . . occur simultaneously (with Ni-H bond lengths identical to within 0.1 Å). Some recent model calculations agree with this insofar as they predict that the two three-fold sites and the top site have higher binding energies than other adsorption sites, with the top site possibly less favorable from the point of view of diffusion. On the other hand, hcp Ti(0001) + p(1 × 1)Cd was found to have the deviating sequence cABABAB . . . , meaning that the cadmium atom is repelled by second-nearest Ti neighbors.

Multilayers of Cd on Ti(0001) have been studied as well, indicating a Cd crystal growth according to the sequence . . acacABAB . . . : the Cd film has the expected hcp structure known for the bulk material. In this case the Ti and Cd lattice constants are sufficiently close to allow growth of the film in registry with the substrate mesh.

A further question regarding the adsorption registry is whether it depends on adsorption coverage, i.e., on density of adatoms: this is relevant to the effects of adatom-adatom interactions. The situation is illustrated by a limited set of results, namely those for quarter-monolayer and half-monolayer adsorption of O, S, Se and Te on Ni(100) in p(2 × 2) and c(2 × 2) periodicities: the adsorption site is found not to depend on coverage in these cases (the nearest adatom-adatom distances are 4.90 and 3.46 Å for the two coverages, respectively, compared with the largest adatom diameter of about 2.7 Å for Te).

Adsorption in many different adsorption sites simultaneously is expected for overlayers with an incommensurate lattice (cf. Sect. III). This has been confirmed by LEED intensity analyses for the case of an incommensurate overlayer of Xe on Ag(111), where both the substrate and the overlayer consist of hexagonally close-packed layers (with unrelated unit cells) parallel to the surface.

Concerning adsorption bond lengths, it is necessary to first recall the uncertainty in the determination of these. Depending on the case (such as on the orientation of the bonds, among other factors) the uncertainty in bond lengths in atomic adsorption determined by LEED varies from about 0.04 about 0.09 Å, corresponding to relative uncertainties of 2% to 4% of the bond length. Medium- and high-energy ion scattering and SEXAFS may have uncertainties of about 0.02 Å or 1% in the few cases examined so far.

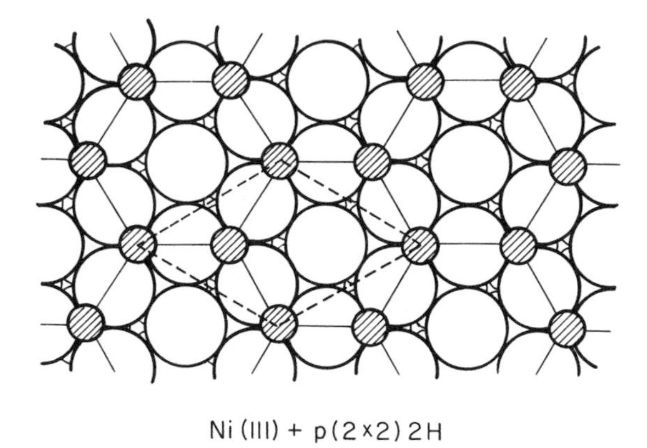

Fig. 6.7. Top view of the Ni(111) + p(2 × 2)2H surface. The dashed lines indicate the unit cell. Adsorbate atoms are drawn shaded

Figure 6.8 reproduces the adsorption bond length information contained in Table 6.1. The first observation is that the bond lengths found at surfaces agree well (with a few exceptions) with those found in other environments: molecules and solid compounds containing the atom pairs under consideration. As is well known[119)] bond

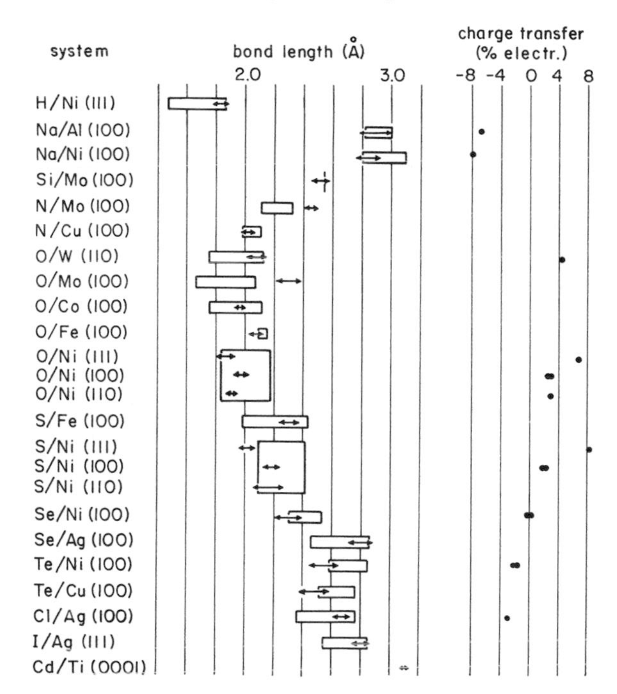

Fig. 6.8. *At left*: comparison of adsorption bond lengths at surfaces (arrows showing uncertainty) with equivalent bond lengths in molecules and bulk compounds (blocks extending over range of values found in standard tables). *At right*: induced charge transfers (obtained as discussed in text) for adsorption

lengths in molecules tend to be smaller than those in bulk compounds, because of the difference in coordination number (number of nearest neighbors). On the whole it seems that bond lengths at surfaces lie closer to the bulk values than the molecular values, which, again on the basis of coordination numbers, seems reasonable.

The uncertainty in the surface bond lengths is sufficiently small that the main bonding mechanisms can probably be investigated. For example, a partial study[132] of systematics in these results, in the spirit of Ref. [119] suggests that the long-established concepts of bond order, valency saturation and resonating bonds are applicable. Model calculations are also beginning to shed light on the particulars of the chemisorption mechanisms, cf. Ref. [32] of Sect. IV.

A non-structural quantity useful in the understanding of chemisorption is the charge transfer between adsorbates and substrates. This charge transfer (giving rise to dipole moments that influence the work function) can be roughly estimated from the observed work function change $\Delta\phi$ during adsorption and the relative positions of the surface atoms, using the relation $\Delta\phi = 4\pi e\sigma d$ for the potential change through a dipole layer, where e is the electronic charge and σ the dipolar charge density; d represents the length of the dipoles and may for example be taken to be the component of the adsorbate-substrate bond length perpendicular to the surface. One may ask now, within this simple model, what fraction $\Delta e/e$ of an electron transferred at each adsorbate site through a distance equal to this component of the bond length produces the measured work function change. This fraction $\Delta e/e$ is plotted in Fig. 6.8, in those cases where work function changes have been measured. The first observation is that $|\Delta e/e|$ is relatively small – at most about 11% – even for alkali adsorbates (which produce the largest work function charges; the larger bond lengths for alkali adsorption probably explain the larger work function changes). A number of model calculations for single adsorbates (the low-coverage limit) also predict such small charge transfers Additionally, at higher coverages one may invoke the effect of dipole-dipole interactions: these tend to reduce the dipole strengths. This last point is confirmed by the observed coverage dependence of the implied charge transfers: the charge transfer per adatom is reduced when the adsorption coverage is increased. However, the behavior of charge transfers is seen to be even more complicated, when one notices that the work function change and the charge transfer can switch their signs on a variation of the coverage, as happens with Se adsorbed on Ni(100)[133], even though the bonding geometry is not noticeably affected. Furthermore, Te on Ni(100) produces charge transfers opposite in sign to O and S on Ni(100), despite the same valency[133].

References

1a. Boudreaux, D. S., Hoffstein, V.: Phys. Rev. *B3*, 2447 (1971);
b. Jepsen, D. W., Marcus, P. M., Jona, F.: Phys. Rev. *B6*, 3684 (1972);
c. Laramore, G. E., Duke, C. B.: Phys. Rev. *B5*, 267 (1972);
d. Martin, M. R., Somorjai, G. A.: Phys. Rev. *B7*, 3607 (1973);
e. Adams, D. L., Landman, U.: Phys. Rev. *B15*, 3775 (1977)
2a. Forstmann, F.; Jap. J. of Appl. Phys., Suppl. 2, Part 2, 657 (1974);

. b. Soria, F., Sacédon, J. L., Echenique, P. M., Titterington, D.: Surf. Sci. *68*, 448 (1977)
3. Ngoc, T. C., Lagally, M. G., Webb, M. B.: Surf. Sci. *35*, 117 (1973)
4. Forstmann, F., Berndt, W., Büttner, P.: Phys. Rev. Lett. *30*, 17 (1973)
5. Head, J. D., Mitchell, K. A. R., Noodleman, L.: Surf. Sci. *61*, 661 (1977)
6. Citrin, P. H., Eisenberger, P., Hewitt, R. C.: J. Vac. Sci. Technol. *15*, 449 (1978)
7a. Cohen, P. I., Unguris, J., Webb, M. B.: Surf. Sci. *58*, 429 (1976);
b. Webb, M. B., Cohen, P. I.: CRC Solid State Sci. *6*, 253 (1976)
8. Stoner, N., Van Hove, M. A., Tong, S. Y., Webb, M. B.: Phys. Rev. Lett. *40*, 243 (1978)
9. Lee, B. W., Alsenz, R., Ignatiev, A., Van Hove, M. A.: Phys. Rev. *B17*, 1510 (1978)
10a. Laramore, G. E.: Phys. Rev., *B9*, 1204 (1974);
b. Watson, P. R., Shepherd, F. R., Frost, D. C., Mitchell, K. A. R.: Surf. Sci. *72*, 562 (1978)
11. Chan, C.-M., Cunningham, S. L., Van Hove, M. A., Weinberg, W. H., Withrow, S. P.: Surf. Sci. *66*, 394 (1977)
12a. Demuth, J. E., Marcus, P. M., Jepsen, D. W.: Phys. Rev. *B11*, 1460 (1975);
b. Feder, R.: Phys. Rev. *B15*, 1751 (1977)
c. Laramore, G. E.: Phys. Rev. *B8*, 515 (1973)
13. Ngoc, T. C., Lagally, M. G., Webb, M. B.: Surf. Sci. *35*, 117 (1973)
14. Christmann, K., Behm, R. J., Van Hove, M. A., Ertl, G., Weinberg, W. H.: J. Chem. Phys. *70*, 4168 (1979)
15. Wang, S. W.: to be published
16. Marcus, P. M., Demuth, J. E., Jepsen, D. W.: Surf. Sci. *53*, 501 (1975)
17a. Demuth, J. E., Jepsen, D. W., Marcus, P. M.: Phys. Rev. Lett. *32*, 1182 (1974)
b. Marcus, P. M., Demuth, J. E., Jepsen, D. W.: Surf. Sci. *53*, 501 (1975)
18. Anderson, A. B.: J. Chem. Phys. *66*, 2173 (1977)
19a. Kesmodel, L. L., Somorjai, G. A.: Phys. Rev. *B11*, 630 (1975)
b. Kesmodel, L. L., Stair, P. C., Somorjai, G. A.: Surf. Sci. *64*, 342 (1977)
20a. Bogh, E., Stensgaard, I.: Proc. 7th IVC and 3rd ICSS, Vienna (1977) p. A-2757
b. Van der Veen, J. F., Smeenk, R. G., Saris, F. W.: Proc. 7th IVC and 3rd ICSS, Vienna (1977), p. 2515
21. Frost, D. C., Mitchell, K. A. R., Shepherd, F. R., Watson, P. R.: Proc. 7th IVC and 3rd ICSS, Vienna (1977), p. A-2725
22. Strozier, J. A., Jones, R. O.: Phys. Rev. *B3*, 3228 (1971)
23. Shih, H. D., Jona, F., Jepsen, D. W., Marcus, P. M.: Commun. on Physics *1*, 25 (1976)
24. Shih, H. D., Jona, F., Jepsen, D. W., Marcus, P. M.: J. Phys. *C9*, 1405 (1976)
25. Shih, H. D., Jona, F., Jepsen, D. W., Marcus, P. M.: Phys. Rev. *B15*, 5550 and 5561 (1977)
26a. Shih, H. D., Jona, F., Jepsen, D. W., Marcus, P. M.: Phys. Rev. Lett. *36*, 798 (1976)
b. Shih, H. D., Jona, F., Jepsen, D. W., Marcus, P. M.: Surf. Sci. *60*, 445 (1976)
27. Unertl, W. N., Thapliyal, H. V.: J. Vac. Sci. Technol. *12*, 263 (1975)
28. Feder, R., Gafner, G.: Surf. Sci. *57*, 45 (1976)
29a. Andersson, S., Pendry, J. B., Echenique, P. M.: Surf. Sci. *65*, 539 (1977)
b. Echenique, P. M.: J. Phys. *C9*, 3193 (1976)
30. Lagally, M. G., Buchholz, J. C., Wang, G. C.: J. Vac. Sci. Technol. *12*, 213 (1975)
31a. Feder, R.: Phys. Stat. Solidi, *b62*, 135 (1974)
b. Van Hove, M. A., Tong, S. Y.: Surf. Sci. *54*, 91 (1976)
32a. Van Hove, M. A., Tong, S. Y.: Phys. Rev. Lett. *35*, 1092 (1975)
b. Van Hove, M. A., Tong, S. Y., Elconin, M. H.: Surf. Sci. *64*, 85 (1977)
33. Tait, R. H., Tong, S. Y., Rhodin, T. N.: Phys. Rev. Lett. *28*, 553 (1972)
34. Masud, N., Kinniburgh, C. G., Pendry, J. B.: J. Phys. *C10*, 1 (1977)
35a. Hutchins, B. A., Rhodin, T. N., Demuth, J. E.: Surf. Sci. *54*, 419 (1976)
b. Van Hove, M. A., Tong, S. Y., Stoner, N.: Surf. Sci. *54*, 259 (1976)
36a. Jepsen, D. W., Marcus, P. M., Jona, F.: Phys. Rev. *B8*, 5523 (1973)
b. Moritz, W.: Doctoral Thesis (University of Munich, 1976)
37. Zanazzi, E., Jona, F., Jepsen, D. W., Marcus, P. M.; Phys. Rev. *B14*, 432 (1976)
38. Ignatiev, A., Jona, F., Jepsen, D. W., Marcus, P. M.: Surf. Sci. *40*, 439 (1973)

39. Feder, R.: Surf. Sci. *68*, 229 (1977)
40. Maglietta, M., Zanazzi, E., Jona, F.: Bull. Am. Phys. Soc. *22*, 355 (1977)
41. Maglietta, M., Zanazzi, E., Bardi, U., Jona, F., Jepsen, D. W., Marcus, P. M.: Surf. Sci. *77*, 101 (1978)
42a. Capart, G.: Surf. Sci. *26*, 429 (1971)
b. Marcus, P. M., Jepsen, D. W., Jona, F.: Surf. Sci. *31*, 180 (1972)
c. Pendry, J. B.: J. Phys. *C4*, 2514 (1971)
43. Pendry, J. B.: Low Energy Electron Diffraction (Academic Press, London, 1974).
44. Kleiman, G. G., Burkstrand, J. M.: Surf. Sci. *50*, 493 (1975)
45. Burkstrand, J. M., Kleiman, G. G., Tibbetts, G. G., Tracy, J. C.: J. Vac. Sci. Technol. *13*, 291 (1976)
46. Burkstrand, J. M., Tong, S. Y., Van Hove, M. A.: to be published
47. Salwén, A., Rundgren, J.: Surf. Sci. *53*, 523 (1975)
48. Tait, R. H., Tong, S. Y., Rhodin, T. N.: Phys. Rev. Lett. *28*, 553 (1972)
49. Van Hove, M. A., Tong, S. Y.: Surf. Sci. *52*, 673 (1975)
50a. Andersson, S., Pendry, J. B.: J. Phys. *C6*, 601 (1973)
b. Andersson, S., Pendry, J. B.: Solid St. Commun. *16*, 563 (1975)
51. Demuth, J. E., Jepsen, D. W., Marcus, P. M.: J. Phys. *C8*, L25 (1975)
52. Demuth, J. E., Jepsen, D. W., Marcus, P. M.: Phys. Rev. Lett. *31*, 540 (1973)
53. Brongersma, H. H., Theeten, J. B.: Surf. Sci. *54*, 519 (1976)
54. Hagstrum, H. D., Becker, G. E.: J. Chem. Phys. *54*, 1015 (1971)
55. Van Hove, M., Tong, S. Y.: J. Vac. Sci. Technol. *12*, 230 (1975)
56. Walch, S. P., Goddard III, W. A.: Sol. St. Comm. *23*, 907 (1977); Surf. Sci. *72*, 645 (1978); Surf. Sci. *75*, 609 (1978)
57. Li. C. H., Connolly, J. W. D.: Surf. Sci. *65*, 700 (1977)
58. Groupe d'Etude des Surfaces, Surf. Sci. *48*, 577 (1975)
59. Theeten, J. B., Brongersma, H. H.: Rev. Phys. Appl. *11*, 57 (1976)
60. Hagstrum, H. D., Becker, G. E.: J. Vac. Sci. Technol. *14*, 369 (1977)
61a. Demuth, J. E., Jepsen, D. W., Marcus, P. M.: J. Phys. *C6*, L307 (1973)
b. Demuth, J. E., Marcus, P. M., Jepsen, D. W.: Phys. Rev. Lett. *32*, 1182 (1974)
62. Feder, R.: Surf. Sci. *68*, 229 (1977)
63. Mitchell, K. A. R., Shepherd, F. R., Watson, P. R., Frost, D. C.: Surf. Sci. *72*, 562 (1978)
64. Ignatiev, A., Pendry, J. B., Rhodin, T. N.: Phys. Rev. Lett. *26*, 189 (1971)
65a. Feder, R.: Phys. Stat. Sol. *58*, K137 (1973)
b. Legg, K. O., Jona, F., Jepsen, D. W., Marcus, P. M.: J. Phys. *C10*, 937 (1977)
66a. Legg, K. O., Jona, F., Jepsen, D. W., Marcus, P. M.: J. Phys. *C8*, L492 (1975)
b. Legg, K. O., Jona, F., Jepsen, D. W., Marcus, P. M.: Phys. Rev. *B16*, 5271 (1977)
67. Anderson, A. B.: Phys. Rev. *B16*, 900 (1977)
68a. Feder, R., Viefhaus, H.: to be published.
b. Legg, K. O., Jona, F., Jepsen, D. W., Marcus, P. M.: Surf. Sci. *66*, 25 (1977)
69a. Clarke, L. J.: Proc. 7th IVC and 3rd ICSS, Vienna,(1977), p. A-2725
b. Felter, T. E., Barker, R. A., Estrup, P. J.: Phys. Rev. Lett. *38*, 1138 (1977)
c. Ignatiev, A., Jona, F., Shih, H. D., Jepsen, D. W., Marcus, P. M.: Phys. Rev. *B11*, 4787 (1975)
d. Noguera, C., Spanjaard, D., Jepsen, D., Ballu, Y., Guillot, C.: J. Lecante, J. Paigne, Y. Petroff, Pinchaux, R., Thiry and R. Cinti, Phys. Rev. Lett. *38*, 1171 (1977)
70. Ignatiev, A., Jona, F., Jepsen, D. W., Marcus, P. M.: Surf. Sci. *49*, 189 (1975)
71. Clarke, L. J.: Proc. 7th IVC and 3rd ICSS, Vienna (1977), p. A-2725
72. Ignatiev, A., Jona, F., Jepsen, D. W., Marcus, P. M.: Phys. Rev. *B11*, 4780 (1975)
73a. Debe, M. K., King, D. A., Marsh, F. S.: Surf. Sci. *68*, 437 (1977)
b. Feder, R.: Phys. Rev. Lett. *36*, 598 (1976)
c. Feder, R.: Surf. Sci. *63*, 283 (1977)
d. Kirschner, J., Feder, R.: Verh. Deutsch. Physik. Ges. *2*, 557 (1978)
e. Lee, B. W., Ignatiev, A., Tong, S. Y., Van Hove, M.: J. Vac. Sci. Technol. *14*, 291 (1977)

74. Feldman, L. C., Kauffman, R. L., Silverman, P. J., Zuhr, R. A., Barrett, J. H.: Phys. Rev. Lett. *39*, 38 (1977)
75a. Barker, R. A., Estrup, P. J., Jona, F., Marcus, P. M.: Sol. St. Comm. *25*, 375 (1978)
b. Debe, M. K., King, D. A.: Phys. Rev. Lett. *39*, 708 (1977)
c. Debe, M. K., King, D. A.: J. Phys. *C10*, L303 (1977)
d. Felter, T. E., Barker, R. A., Estrup, P. J.: Phys. Rev. Lett. *38*, 1138 (1977)
76. Ignatiev, A., Lee, B. W., Van Hove, M. A.: to be published
77. Chan, C.-M., Cunningham, S. L., Van Hove, M. A., Weinberg, W. H.: Surf. Sci. *67*, 1 (1977)
78. Finnis, M. W., Heine, V.: J. Phys. *F4*, L37 (1974)
79. Maglietta, M., Zanazzi, E., Jona, F., Jepsen, D. W., Marcus, P. M.: J. Phys. *C10*, 3287 (1977)
80. Zanazzi, E., Maglietta, M., Bardi, U., Jona, F., Jepsen, D. W., Marcus, P. M.: Proc. 7th IVC and 3rd ICSS, Vienna (1977) p. 2447
81. Heiland, W., Iberl, F., Taglauer, E., Menzel, D.: Surf. Sci. *53*, 383 (1975)
82. Davis, H. L.: to be published
83. Chan, C.-M., Van Hove, M. A., Weinberg, W. H., Williams, E. D., to be published
84. Chan, C.-M., Luke, K. L., Van Hove, M. A., Weinberg, W. H., J. Vac. Sci. Techn. *16*, 642 (1979)
85. Chan, C.-M., Luke, K. L., Van Hove, M. A., Weinberg, W. H., Withrow, S. P.: Surf. Sci. *78*, 386 (1978)
86. Van der Veen, J. F., Smeenk, R. G., Tromp, R. M., Saris, F. W.: Ned. Tijdschr, v. Vacuumtechn. *2/3/4*, 284 (1978)
87. Demuth, J. E.: J. of Colloid and Interface Science *58*, 184 (1977)
88. Marcus, P. M., Demuth, J. E., Jepsen, D. W.: Surf. Sci. *53*, 501 (1975)
89. Heiland, W., Schaffler, H. G., Taglauer, E.: Surf. Sci. *35*, 381 (1973)
90. Brongersma, H. H., Theeten, J. B.: Surf. Sci. *54*, 519 (1976)
91. Van der Veen, J. F., Tromp, R. M., Smeenk, R. G., Saris, F. W.: to be published.
92. Walch, S. P., Goddard III, W. A.: Surf. Sci. *72*, 645 (1979)
93. Frost, D. C., Hengrasmee, S., Mitchell, K. A. R., Shepherd, F. R., Watson, P. R.: Surf. Sci. *76*, L585 (1978)
94. Shih, H. D., Jona, F., Jepsen, D. W., Marcus, P. M.: Bull. Am. Phys. Soc. *22*, 357 (1977)
95. Streater, R. W., Moore, W. T., Watson, P. R., Frost, D. C., Mitchell, K. A. R.: Surf. Sci. *72*, 744 (1978)
96. Ignatiev, A., Lee, B. W., Van Hove, M. A.: Proc. 7th IVC and 3rd ICSS, Vienna (1977), p. 1733
97a. Mrstik, B. J., Van Hove, M. A., Tong, S. Y.: Bull. Am. Phys. Soc. *23*, 391 (1978)
b. Tong, S. Y., Van Hove, M. A., Mrstik, B. J.: Proc. 7th IVC and 3rd ICSS, Vienna (1977), p. 2407
98a. Duke, C. B., Lubinsky, A. R., Lee, B. W., Mark, P.: J. Vac. Sci. Technol. *13*, 761 (1976)
b. Lubinsky, A. R., Duke, C. B., Lee, B. W., Mark, P.: Phys, Rev. Lett. *36*, 1058 (1976)
c. Mark, P., Cisneros, G., Bonn, M., Kahn, A., Duke, C. B., Paton, A., Lubinsky, A. R.: J. Vac. Sci. Technol. *14*, 910 (1977)
d. Tong, S. Y., Lubinsky, A. R., Mrstik, B. J., Van Hove, M. A.: Phys. Rev. *B17*, 3303 (1978)
99a. Rowe, J. E., Christman, S. B., Margaritondo, G.: Phys. Rev. Lett. *35*, 1471 (1975)
b. Spicer, W. E., Chye, P. W., Gregory, P. E., Sukegawa, T., Babaloba, I. A.: J. Vac. Sci. Technol. *13*, 233 (1976)
c. Chelikowsky, J. R., Louie, S. G., Cohen, M. L.: Phys. Rev. *B14*, 4724 (1976)
d. Calandra, C., Manghi, F., Bertoni, C. M.: J. Phys. *C10*, 1911 (1977)
e. Mele, E. J., Joannopoulos, J. D.: Phys. Rev. *B17*, 1816 (1978)
f. Goddard, W. A., III, Barton, J. J., Redondo, A., McGill, T. C.: to be published
g. Chadi, D. J.: Phys. Rev. Lett. *41*, 1062 (1978)
100a. Kinniburgh, C. G.: J. Phys *C8*, 2382 (1975)
b. Kinniburgh, C. G.: J. Phys. *C9*, 2695 (1976)
101a. Mrstik, B. J., Kaplan, R., Reinecke, T. L., Van Hove, M., Tong, S. Y.: Phys. Rev. *B15*, 897 (1977)

b. Mrstik, B. J., Kaplan, R., Reinecke, T. L., Van Hove, M., Tong, S. Y.: Il Nuovo Cimento *38B*, 387 (1977)
102. Andersson, S., Pendry, J. B., Echenique, P. M.: Surf. Sci. *65*, 539 (1977)
103. Kinniburgh, C. G., Walker, J. A.: Surf. Sci. *63*, 274 (1977)
104. Shih, H. D., Jona, F., Jepsen, D. W. Marcus, P. M.: Phys. Rev. Lett. *37*, 1622 (1976)
105a. Auer, P. P., Mönch, W.: to be published
b. Snyder, L. C., Wasserman, Z.: Surf. Sci. 77, 52 (1978)
106a. Verwoerd, W. S., Kok, F. J.: Ned. Tijdschr. v. Vacuumtechn. *2/3/4*, 303 (1978)
b. Chadi, D. J.: Phys. Rev. Lett. *41*, 1062 (1978)
107a. Levine, J. D., McFarlane, S. H., Mark, P.: Phys. Rev. *B16*, 5415 (1977)
b. Mark, P., Levine, J. D., McFarlane, S. H.: Phys. Rev. Lett. *38*, 1408 (1977)
108. Appelbaum, J. A., Hamann, D. R.; Surf. Sci. *74*, 21 (1978)
109a. Mitchell, K. A. R., Van Hove, M. A.: Surf. Sci. *75*, 147L (1978)
b. Tong, S. Y., Maldonado, A. L.: Surf. Sci., *78*, 459 (1978)
110. Tong, S. Y.: private communication
111. Appelbaum, J. A., Baraff, G. A., Hamann, D. R.: Phys. Rev. Lett. *35*, 729 (1975); Phys. Rev. *B14*, 588 (1976)
112. White, S. J., Woodruff, D. P., Holland, B. W., Zimmer, R. S.: Surf. Sci. *68*, 457 (1977)
113. Lau, B., Mrstik, B. J., Tong, S. Y., Van Hove, M. A.: to be published
114. Duke, C. B., Lubinsky, A. R.: Surf. Sci. *50*, 605 (1975)
115. Duke, C. B., Lubinsky, A. R., Chang, S. C., Lee, B. W., Mark, P.: Phys. Rev. *B15*, 4865 (1977)
116. Duke, C. B., Lubinsky, A. R., Lee, B. W., Mark, P.: J. Vac. Sci. Technol. *13*, 761 (1976)
117. Lubinsky, A. R., Duke, C. B., Chang, S. C., Lee, B. W., Mark, P.: J. Vac. Sci. Technol. *13*, 189 (1976)
118a. Duke, C. B., Lubinsky, A. R., Bonn, M., Cisneros, G., Mark, P.: J. Vac. Sci. Technol. *14*, 294 (1977)
b. Mark, P., Cisneros, G., Bonn, M., Kahn, A., Duke, C. B., Paton, A., Lubinsky, A. R.: J. Vac. Sci. Technol. *14*, 910 (1977)
119. Pauling, L.: "The Nature of the Chemical Bond," 3rd edition, Cornell Univ. Press, 1960
120. Wasserman, H. J., Vermaak, J. S.: Surf. Sci. *32*, 168 (1972)
121. Van Hove, M. A., Echenique, P. M.: Surf. Sci. *82*, L298 (1978)
122. See, for example, Messmer, R. P., Salahub, D. R.: Phys. Rev. *B16*, 3415 (1977) and references therein
123. Hagstrom, S., Lyon, H. B., Somorjai, G. A.: Phys. Rev. Lett. *15*, 491 (1965)
124a. Grant, J. T.: Surf. Sci. *18*, 282 (1969)
b. Ignatiev, A., Jones, A. V., Rhodin, T. N.: Surf. Sci. *30*, 573 (1972)
125a. Fedak, D. G., Gjostein, N. A.: Surf. Sci. *8*, 77 (1967)
b. Palmberg, P. W., Rhodin, T. N.: Phys. Rev. *161*, 586 (1967)
126. Weinberg, W. H.: private communication.
127. Moritz, W., Wolf, D.: to be published
128a. Zehner, D. M., Appleton, B. R., Noggle, T. S., Miller, J. W., Barrett, J. H., Jenkins, L. H., Schorr, O. E., III: J. Vac. Sci. Technol. *12*, 454 (1975)
b. Wolf. D., Zimmer, A.: private communication
129a. Inglesfield, J. E.: J. Phys. *C11*, L69 (1978)
b. Tosatti, E.: Sol. St. Comm. *25*, 637 (1978)
130. Felter, T. E., Barker, R. A., Estrup, P. J.: Phys. Rev. Lett. *38*, 1138 (1977)
131. Schlier, R. E., Farnsworth, H. E.: "Semiconductor Surface Physics," Univ. of Pennsylvania Press, Philadelphia (1975), p. 3
132. Madhukar, A.: Sol. St. Comm. *16*, 461 (1975)
133. Demuth, J. E., Rhodin, T. N.: Surf. Sci. *45*, 249 (1974)

VII Surface Crystallography of Ordered Multi-Atomic and Molecular Monolayers

1 Introduction

Molecules deposited on surfaces may retain their basic molecular character, bonding as a whole lightly to the substrate. They may dissociate into their constituent atoms, which then bond individually to the substrate. Molecules may instead break up into smaller fragments which become largely independent or recombine into other configurations. Intermediate cases also can occur, such as with relatively strong bonding of molecules with resulting strong distortions. In addition, countless cases of co-adsorption of different atoms or different molecules can be investigated. The exact form that such multi-atomic or molecular adsorption takes among those mentioned above is known to depend strongly on the temperature, as will be discussed below: this provides a way to determine the bond strengths.

The study of the structure of multi-atomic and molecular adsorbates is in its very early stages, but more and more efforts are devoted to it because of its obvious importance in catalysis and other fields. The techniques of investigation used are primarily LEED, photoemission and high resolution electron energy loss spectroscopy. The detailed adsorption geometry has been analyzed so far for a few cases of co-adsorption of atoms and a few adsorbed molecules. The ordering characteristics of molecular monolayers have been investigated for a number of small molecules and a sizeable family of larger organic molecules.

2 Co-Adsorption of Atoms

One small family of surfaces created by coadsorption of two different atomic species has been structurally investigated[1]. The substrate is fcc Ni(100) which is not structurally affected by the adsorption. The adatoms are S and Na, deposited sequentially in that order, each in either half-coverage c(2 × 2) or quarter-coverage p(2 × 2) ordered overlayers. With a half monolayer of each species (cf. Fig. 7.1a), the position of the S atoms in hollow sites is not affected by the addition of the Na atoms; the Na atoms choose the unoccupied hollow sites on the substrate, where they have 4 nearest S neighbors with a Na-S bond length of 2.76 ± 0.1 Å (compared with 2.735–3.38 in a number of bulk compounds). The Na atoms are 0.2 Å farther away from the substrate than in the absence of S, an increase by 0.15 Å of the Ni-Na bond length. Halving the

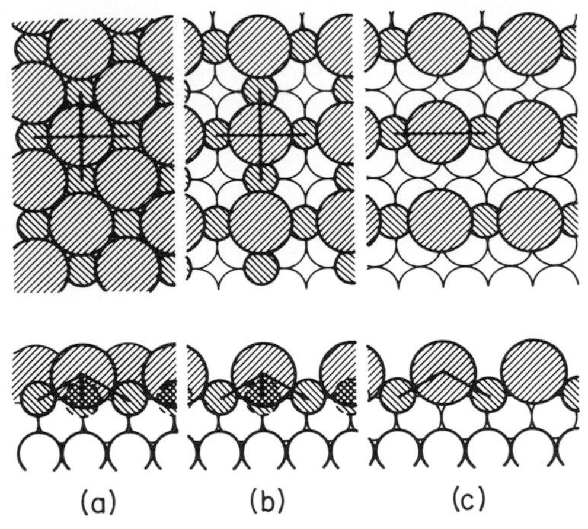

Fig. 7.1 a–c. The co-adsorption geometry of S (*small shaded circles*) and Na (*large shaded circles*) on Ni(100) (*open circles*), in top and side views: (**a**) half-monolayer of S and half-monolayer of Na; (**b**) half-monolayer of S and quarter-monolayer of Na; (**c**) quarter-monolayer of S and quarter-monolayer of Na

Na coverage, leaving that of S unchanged, does not affect these results (cf. Fig. 7.1b), indicating little charge effect in the bonding. This last impression is confirmed by work function measurements. For half-monolayer coverage of both species the work function change relative to the bare substrate is –2.65 eV (compared with –2.55 eV in the absence of S), while halving the Na coverage yields –2.85 eV; this halving therefore induces a charge transfer of the order of 1% of an electron between two atoms.

With a quarter monolayer of both S and Na, again the position of the S atoms is insensitive to the addition of Na atoms (cf. Fig. 7.1c), and again the Na atoms choose unoccupied hollow sites on the substrate, but only those sites that provide the closest contact with S atoms, rather than the sites that allow closer contact with the substrate. So an attractive force acts between the coadsorbed species. Again the Na–S bond length is 2.76 ± 0.1 Å, even though the number of nearest S atoms is now reduced from 4 to 2. The work function change (relative to the bare substrate) is now –3.10 eV so that again little charge effect is seen in the bonding, despite the fact that Na and S could be expected to have strong ionic character. The mutual destruction of parallel dipoles seems to play a significant role here, as in all cases of high-coverage adsorption discussed previously.

3 Dissociative Adsorption of Carbon Monoxide

On many metal substrates, CO dissociates into individual atoms that in some cases still produce an ordered monolayer. The resulting geometry has been investigated for a titanium substrate and for an iron substrate.

With Ti(0001) + p(2 × 2) CO, preliminary results[2)] suggest that the C and O atoms occupy threefold hollow sites, the C atoms forming a p(2 × 2) array and the O atoms forming a similar but shifted p(2 X 2) array. Both C and O are found to probably choose the same type of hollow site, but which of the two inequivalent threefold sites (bABC . . . or cABC . . .) is not known.

In the case of Fe(100) + c(2 X 2)CO, the LEED analysis[3] finds that the C and O atoms individually and randomly occupy fourfold hollow sites in a c(2 X 2) array, i.e., a c(2 X 2) array of unoccupied sites exists, all other sites being occupied at random by either C or O atoms. The average Fe–C and Fe–O bond length is 1.93 Å (C and O usually have very similar radii), somewhat smaller than for Fe(100) + p(1 X 1)O (where it is about 2.08 Å); however, an expansion of the topmost substrate interlayer spacing has not been considered in this dissociative case (the bulk spacing was assumed), resulting in some uncertainty in the Fe-adsorbate bond length as well.

4 Adsorption Geometry of Molecules

Determinations of the surface structure by computing the diffraction beam intensities from low energy electron diffraction are concentrated in two frontier areas at present. One is the determination of the surface structures of adsorbed molecules of ever bigger size and the other is the determination of the atomic locations in reconstructed clean solid surfaces.

So far, only a very few adsorbed molecular structures have been analyzed by surface crystallography. The first system studied in detail was acetylene adsorbed on the (111) crystal face of platinum. We shall discuss the complex adsorption and structural characteristics of this small organic molecule in some detail as it reveals the unique surface bonding arrangements that are possible and points to the importance of the use of additional techniques to complement the diffraction information.

Acetylene forms spontaneously an ordered (2 X 2) surface structure on the Pt(111) surface at 300 K, at low exposure under ultrahigh vacuum conditions. The intensity profiles reveal that this structure is metastable, and upon heating to 350–400 K for one hour, it undergoes a transformation to a stable structure with the same (2 X 2) unit cell. Ethylene adsorbs on the Pt(111) surface and at 300 K, it forms an ordered (2 X 2) surface structure that is identical to the stable acetylene structure as shown by the intensity profiles.

Data was taken in the electron energy range of 10–200 eV, but little sensitivity to the organic adsorbate is found above ~100 eV. The observed diffraction pattern arises from three equivalent 120° – rotated domains of (2 X 2) unit cells. The optimum agreement between calculated and experimental intensity data for the metastable acetylene structure is achieved for an atop site coordination[4]. The molecule is located at a z-distance of 2.5 Å from the underlying surface platinum atom. However, the best agreement is obtained if the molecule is moved toward a triangular site, where there is a platinum atom in the second layer, by 0.25 Å, as shown in Fig. 7.2.

The same system, i.e., C_2H_2 on Pt(111) has also been studied by ultraviolet photoelectron spectroscopy (UPS), by high resolution electron energy loss spectroscopy (HREELS), and by thermal desorption spectroscopy (TDS). The authors have all

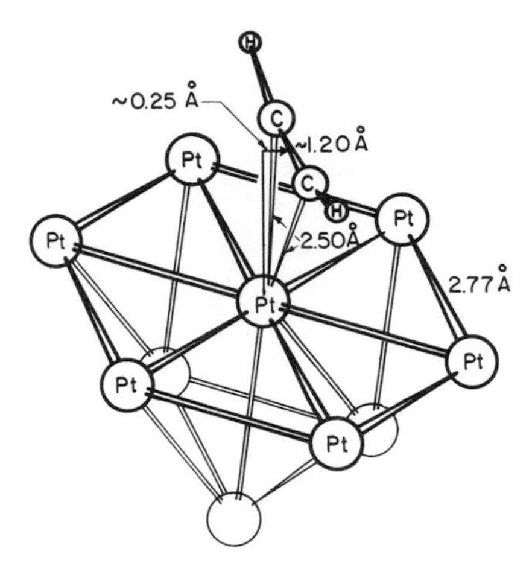

Fig. 7.2. Perspective view of metastable C_2H_2 on Pt(111) (hydrogen atom positions are uncertain). Thin-lined atoms belong to second substrate layer

reported the presence of at least two states of binding and conversion from one state to the other as a function of temperature.

To solve the stable acetylene surface structure the combined experimental informations that came from LEED and ELS were required[5]. ELS studies of the vibrational spectrum indicated the presence of a methyl group and that the molecule must be lined up at some angle to the crystal surface. LEED structure analysis determined that the species is coordinated to a threefold site where there is no metal atoms underneath, in the second layer. The C-C axis is normal to the surface within our uncertainty of 15°. The C–C bond length was found to be 1.50 Å ± 0.05 Å. This value is nearly identical to the single bond carbon-carbon distance in most saturated organic molecules. There are also three equivalent Pt–C bond lengths of 2.00 Å ± 0.05 Å. The surface species most consistent with all of the studies is ethylidyne ($\geqslant C{-}CH_3$) and its structure is shown in Fig. 7.3. This structure is found also in many organometallic acetylene clusters. The ethylidyne group forms readily upon exposure of ethylene (C_2H_4) to the Pt(111) surface with the transfer of one hydrogen atom to the surface per ethylene.

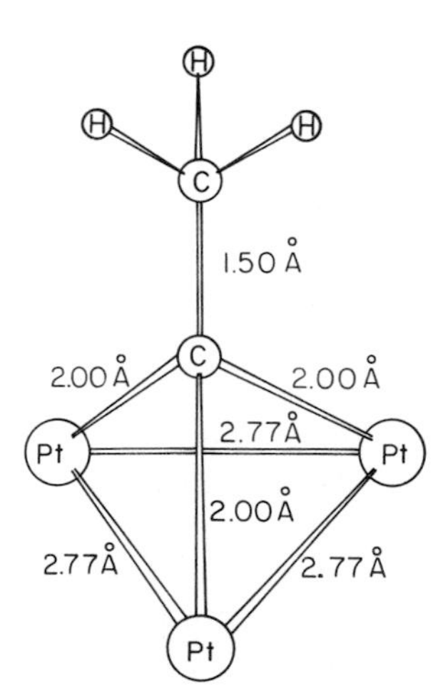

Fig. 7.3. Perspective view of ethylidyne on Pt(111), the stable structure reached after acetylene adsorption with hydrogen addition

The complete conversion of C_2H_2 to ethylidyne requires the presence of surface hydrogen atoms and proceeds rapidly only at ~350 K. By comparison with reported reaction mechanisms on related transition metal clusters it seems likely that vinylidene ($>C = CH_2$) is an intermediate during the conversion from the metastable to the stable acetylene structure. An interesting question is the source of hydrogen that must be attached to the molecule to form ethylidyne from acetylene. It appears that there is enough hydrogen on the metal surface from the residual background of the ultrahigh vacuum diffraction chamber to provide the hydrogen necessary for the 1/4 monolayer of adsorbate. The disproportionation reaction of C_2H_2 can be ruled out as neither LEED nor HREELS show evidence for the presence of more than one surface species. Indeed, increased hydrogen partial pressures during the adsorption studies facilitates the formation of the stable surface structure of acetylene from the metastable structure.

Both the long C–C bond distance (1.50 Å) and the very short Pt–C distances (2.0 Å) indicate the strong interaction between the adsorbed molecule and the three platinum surface atoms. The covalent Pt–C distance would be 2.2 Å. The shorter metal-carbon distances indicate multiple metal-carbon bonding that may be carbene or carbyne-like. Compounds with these types of bonds exhibit high reactivity in metathesis and in other addition reactions[6]. The carbon-carbon single bond distance indicates that the molecule is stretched as much as possible without breaking of this chemical bond.

It is likely that the unique surface and catalytic chemistry of platinum is associated with the formation of hydrocarbon molecular intermediates of the type produced by the adsorption of C_2H_2 or C_2H_4. Metals to the left of platinum in the periodic table would form stronger metal-carbon bonds. As a result the carbon-carbon bond would snap and molecular fragments would form instead of the ethylidyne species. LEED and HREELS studies of the structure and bonding of C_2H_2 and C_2H_4 on Ni and Pd(111) surfaces are in progress. Preliminary electron spectroscopy evidence indicates that the molecules remain oriented parallel to these metal surfaces.

The second-molecular system that has recently been studied is CO in a c(2 × 2) arrangement on the Ni(100) crystal face[7]. It appears from LEED that this molecule is bound by its carbon end to one nickel atom with a Ni–C bond length of 1.8 ± 0.1 Å, cf. Fig. 7.4. The carbon-end bonding configuration has long been expected from UPS and IR evidence and HREELS confirms bonding to a single nickel atom. However, the CO internuclear axis is observed not to be perpendicular to the surface but tilted by 34 ± 10° from the surface normal. But photoemission results do favor a perpendicular position of the molecule.

A LEED analysis of CO adsorbed in a $(2\sqrt{2} \times \sqrt{2})$ arrangement on Pd(100) also indicates molecular adsorption[8], as expected from previous studies. However there is bridge-bonding of the carbon ends to pairs of metal atoms with a Pd–C bond length of 1.90 ± 0.06 Å and no noticeable tilting of the CO axis from the perpendicular to the surface, the CO bond length being 1.15 ± 0.1 Å, cf. Fig. 7.5. There is some indication of relatively large vibration amplitudes for the O atoms, the nature of these vibrations remaining unresolved.

Comparing these results for CO bonded to Ni(100) and Pd(100) to the structure of metal carbonyl clusters, one finds that the multiply-coordinated CO on palladium has

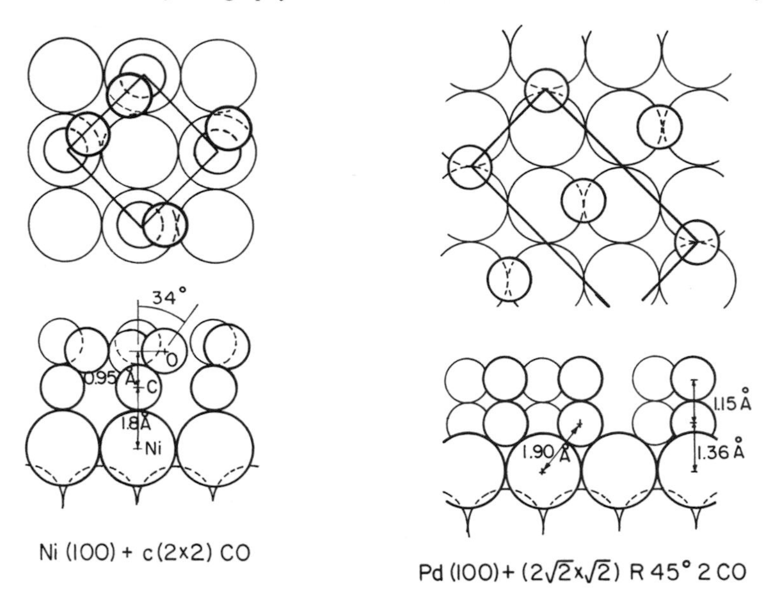

Fig. 7.4. Proposed structure of CO on Ni(100) from LEED studies, in top and side views. The CO molecules are tilted 34° away from the surface normal; the unknown tilt azimuth is chosen random here for illustration purposes. The C and O atoms are given equal touching-sphere radii

Fig. 7.5. The adsorption structure of CO on Pd(100) at a half-monolayer coverage, in top and side views

relatively smaller metal-carbon bond lengths than in terminal-bonding to nickel, suggesting a stronger bonding to palladium. However the heats of adsorption are rather similar.

Note Added in Proof. New LEED analyses show that CO on Ni(100) is perpendicular to the surface.

References

1. Andersson, S., Pendry, J. B.: J. Phys. *C9*, 2721 (1976)
2. Shih, H. D., Jona, F., Jepsen, D. W., Marcus, P. M.; J. Vac. Sci. Technol. *15*, 596 (1978)
3. Jona, F., Legg, K. O., Shih, H. D., Jepsen, D. W., Marcus, P. M.: Phys. Rev. Lett. *40*, 1466 (1978)
4. Kesmodel, L. L., Baetzold, R., Somorjai, G. A.: Surf. Sci. *66*, 299 (1977)
5. Kesmodel, L. L., Dubois, L. H., Somorjai, G. A.: Chem. Phys. Lett. *56*, 267 (1978)
6. Fischer, E. O., Massböl, A.: Angew. Chem. *76*, 345 (1964); Schrock, R. R.: J. Am. Chem. Soc. *98*, 5399 (1976)
7. Andersson, S., Pendry, J. B.: Surf. Sci. *71*, 75 (1978)
8. Behm, H. J., Christmann, K., Ertl, G., Van Hove, M. A.: to be published

Acknowledgement. One of us (MAVH) wishes to acknowledge the support by Sonderforschungsbereich 128 (Deutsche Forschungsgemeinschaft) during part of this work.

We are grateful to D. Castner for updating the list of adsorbate surface structures and to J. P. Biberian for his tabulation of the surface structures of metals adsorbed on other metal surfaces.

This work was supported by the Division of Materials Sciences, Office of Basic Energy Sciences, U.S. Department of Energy, under Contract No. W-7405-Eng-48.

Subject Index

Author-Index Volumes 1—38

Structure and Bonding

Editors: J. D. Dunitz, J. B. Goodenough, P. Hemmerich, J. A. Ibers, C. K. Jørgensen, J. B. Neilands, D. Reinen, R. J. P. Williams

Volume 31

Bonding and Compounds of Less Abundant Metals

1976. 40 figures, 6 tables. IV, 111 pages
ISBN 3-540-07964-5

Contents:
R. Ferreira: Paradoxial Violations of Koopman's Theorem with Special Reference to the 3 *d* Transition Elements and the Lanthanides. *C. Bonnelle:* Band and Localized States in Metallic Thorium, Uranium, and Plutonium, and in some Compounds, Studied by X-Ray Spectroscopy
V. Gutmann, H. Mayer: The Application of the Functional Approach to Bond Variations under Pressure
J. K. Burdett: The Shapes of Main-Group Molecules. A Simple Semi-Quantitative Molecular Orbital Approach

Volume 32

Novel Chemical Effects of Electronic Behaviour

1977. 31 figures, 16 tables. IV, 171 pages
ISBN 3-540-08014-7

Contents:
H. Vahrenkamp: Recent Results in the Chemistry of Transition Metal Clusters with Organic Ligands
J. A. Wilson: A Generalized Configuration-Dependent Band Model for Lanthanide Compounds and Conditions for Interconfiguration Fluctuations
J. N. Murrell: The Potential Energy Surfaces of Polyatomic Molecules
J. A. Duffy: Optical Electronegativity and Nephelauxetic Effect in Oxide Systems: Application to Conducting, Semi-Conducting and Insulating Metal Oxides

Volume 33

New Concepts

1977. 85 figures, 47 tables. IV, 214 pages
ISBN 3-540-08269-7

Contents:
W. E. Wallace, S. G. Sankar, V. U. S. Rao: Crystal Field Effects in Rare-Earth Intermetallic Compounds
D. K. Hoffmann, R. Ruedenberg, J. G. Verkade: Molecular Orbital Bonding Concepts in Polyatomic Molecules: A Novel Pictorial Approach
K. D. Warren: Ligand Field Theory of *f*-Orbital Sandwich Complexes
K. Schubert: The Two-Correlations Model, a Valence Model for Metallic Phases
C. Linarès, A. Louat, M. Blanchard: Rare-Earth Oxygen Bonding in the $LnMO_4$ Xenotime Structure: Spectroscopic Investigation and Comparative Study of Ligand Field Models

Volume 34

Novel Aspects

1978. 37 figures, 39 tables. IV, 220 pages
ISBN 3-540-08676-5

Contents:
S. Hubert, M. Hussonnois, R. Guillaumont: Measurement of Complexing Constants by Radiochemical Methods
C. K. Jørgensen: Predictable Quarkonium Chemistry
R. A. Bulman: Chemistry of Plutonium and the Transuranics in the Biosphere
J. W. Buchler, W. Kokisch, P. D. Smith: Cis, Trans, and Metal Effects in Transition Metal Porphyrins
D. K. Koppikar, P. V. Sivapullaiah, L. Ramakrishnan, S. Soundarajan: Complexes of the Lanthanides with Neutral Oxygen Donor Ligands

Volume 35

New Theoretical Aspects

1978. 41 figures, 15 tables. IV, 175 pages
ISBN 3-540-08887-3

Contents:
J.-F. Labarre: Conformational Analysis in Inorganic Chemistry: Semi-Empirical Quantum Calculations vs. Experiment
D. B. Cook: The Approximate Calculation of Molecular Electronic Structures as a Theory of Valence
D. W. Smith: Applications of the Angular Overlap Model
C. Furlani, C. Sauletti: He(I) Photoelectron Spectra of *d*-Metal Compounds

Volume 36

Inorganic Chemistry and Spectroscopy

1979. 69 figures, 22 tables. IV, 178 pages
ISBN 3-540-09201-3

Contents:
R. J. H. Clark, B. Stewart: The Resonance Raman Effect
A. Simon: Structure and Bonding with Alkali Metal Suboxides
C. Daul, C. W. Schläpfer, A. von Zelewsky: The Electronic Structure of Cobalt (II) Complexes with Schiff Bases and Related Ligands

Volume 37

Structural Problems

1979. 40 figures, 43 tables. IV, 216 pages
ISBN 3-540-09455-5

Contents:
D. Reinen, C. Friebel: Local and Cooperative Jahn-Teller Interactions in Model Structures. Spectroscopic and Structural Evidence
E. E. Hellner: The Frameworks (Bauverbände) of the Cubic Structure Types
J. Shamir: Polyhalogen Cations

Crystals

Growth, Properties, and Applications

Managing Editor: C. J. M. Rooijmans

Volume 1

Crystals for Magnetic Applications

Editor: C. J. M. Rooijmans

1978. 79 figures, 8 tables. VI, 139 pages
ISBN 3-540-09002-9

Contents:
W. Tolksdorf, F. Welz: Crystal Growth of Magnetic Garnets from High-Temperature Solutions
F. J. Bruni: Gadolinium Gallium Garnet
M. H. Randles: Liquid Phase Epitaxial Growth of Magnetic Garnets
L. N. Demianets: Hydrothermal Crystallization of Magnetic Oxides
M. Sugimoto: Magnetic Spinel Single Crystals by Bridgman Technique.

Volume 2

Growth and Properties

Editor: H. C. Freyhardt

1979. 97 figures, 20 tables. Approx. 290 pages
ISBN 3-540-09697-3

Contents:
K. Nassau, J. Nassau: The Growth of Synthetic and Imitation Gems
E. Schönherr: The Growth of Large Crystals from the Vapor Phase
D. E. Ovsienko, G. A. Alfintsev: Crytal Growth from the Melt/Experimental Investigation of Kinetics and Morphology
A. H. Morrish: Morphology and Physical Properties of Gamma Iron Oxide

Springer-Verlag Berlin Heidelberg New York